# ECOLOGICAL TECHNOLOGIES FOR INDUSTRIAL WASTEWATER MANAGEMENT

## Petrochemicals, Metals, Semi-Conductors, and Paper Industries

# ECOLOGICAL TECHNOLOGIES FOR INDUSTRIAL WASTEWATER MANAGEMENT

## Petrochemicals, Metals, Semi-Conductors, and Paper Industries

*Edited by*
**Victor M. Monsalvo, PhD**

**AAP** APPLE ACADEMIC PRESS

Apple Academic Press Inc. | Apple Academic Press Inc.
3333 Mistwell Crescent | 9 Spinnaker Way
Oakville, ON L6L 0A2 | Waretown, NJ 08758
Canada | USA

©2015 by Apple Academic Press, Inc.

First issued in paperback 2021

*Exclusive worldwide distribution by CRC Press, a member of Taylor & Francis Group*

No claim to original U.S. Government works

ISBN 13: 978-1-77463-559-9 (pbk)
ISBN 13: 978-1-77188-147-0 (hbk)

---

### Library and Archives Canada Cataloguing in Publication

---

Ecological technologies for industrial wastewater management : petrochemicals, metals, semi-conductors, and paper industries / edited by Victor M. Monsalvo.

Includes bibliographical references and index.
ISBN 978-1-77188-147-0 (bound)
1. Sewage--Purification. 2. Ecological engineering. I. Monsalvo, Victor (Victor M.), author, editor

TD745.E36 2015            628.3              C2015-901517-0

---

### Library of Congress Cataloging-in-Publication Data

---

Ecological technologies for industrial wastewater management : petrochemicals, metals, semi-conductors, and paper industries / Victor M. Monsalvo, editor.

pages cm
Includes bibliographical references and index.
ISBN 978-1-77188-147-0 (alk. paper)
1. Industrial water supply. 2. Water quality management. 3. Water conservation. 4. Effluent quality.  I. Araujo Monsalvo, Víctor Manuel, editor.

TD220.2.E36 2015        628.1'683--dc23          2015007540

---

Apple Academic Press also publishes its books in a variety of electronic formats. Some content that appears in print may not be available in electronic format. For information about Apple Academic Press products, visit our website at **www.appleacademicpress.com** and the CRC Press website at **www.crc-press.com**

# ABOUT THE EDITOR

**VICTOR M. MONSALVO**

Professor Victor Monsalvo is an environmental scientist with a PhD in chemical engineering from the University Autonoma de Madrid, where he later became a professor in the chemical engineering section. As a researcher, he has worked with the following universities: Leeds, Cranfield, Sydney, and Aachen. He took part of an active research team working in areas of environmental technologies, water recycling, and advanced water treatment systems. He has been involved in sixteen research projects sponsored by various entities. He has led nine research projects with private companies and an R&D national project, coauthored two patents (national and international) and a book, edited two books, and written around fifty journal and referred conference papers. He has given two key notes in international conferences and has been a member of the organizing committee of five national and international conferences, workshops, and summer schools. He is currently working as senior researcher in the Chemical Processes Department at Abengoa Research, Abengoa.

# CONTENTS

# ACKNOWLEDGMENT AND HOW TO CITE

The editor and publisher thank each of the authors who contributed to this book. The chapters in this book were previously published elsewhere. To cite the work contained in this book and to view the individual permissions, please refer to the citation at the beginning of each chapter. Each chapter was read individually and carefully selected by the editor; the result is a book that provides a comprehensive perspective on industrial wastewater management. The chapters included examine the following topics:

- Chapter 1 assesses the anaerobic biological treatment of wastewater resulting from the regeneration process of used industrial oil by determining the optimal operating conditions of an expanded granular sludge bed reactor.
- Chapter 2 outlines the authors' search for new anionic sorbents, the use of which would allow more effective removal of phenols from wastewater.
- Oily wastewater is a major environmental problem, which conventional water treatments are insufficient to address. Chapter 3 compares the performance of two viable treatment methods: a rotating biological contactor and a hybrid membrane reactor.
- Heavy metals released into the environment via wastewater from steel industries are persistent environmental contaminants that accumulate throughout the food chain with serious ecological and human-health effects. Chapter 4 explores the potential of marine biomass to offer a cheap and efficient elimination technology.
- Chapter 5 investigates the optimal conditions, including effects of iron concentration and electric current, for electrolytic recovery of copper from electroplating wastewater.
- Chapter 6 explores the potential of the yeast *Rhodotorula mucilaginosa* to remove and convert copper ions to copper nanoparticles, a cheap and effective wastewater treatment method.
- The production of semiconductors often uses TMAH (tetramethyl ammonium hydroxide), which is then released into the environment via wastewater. The authors of chapter 7 assess the function of UV light, a magnetic catalyst, and hydrogen peroxide on enhancing ozone in order to mineralize TMAH
- Chapter 8 identifies the characteristics of semiconductor wastewater before and after treatment, the quality that must be maintained in the environment

to which the wastewater is to be discharged or reused, and methods for analyzing various physical and chemical parameters.

- Electroplating is one of the important steps in the production of semiconductors, producing wastewater with high concentration of suspended solids, organic compounds, and dissolved cation. The authors of chapter 9 investigate more efficient polyelectrolytes treatment with low environmental impact.

- In Chapter 10, the authors study ways to improve the biodegradability index of wastewater from pulp and paper industries, using electrocoagulation and different advanced oxidation processes.

- The lignin released into wastewater by paper and pulp industries cannot be easily removed by traditional biochemical treatments. In chapter 11, the authors offer further research into the electrocoagulation technique for paper and pulp wastewater treatment.

- Different types of paper mills produce different types of effluent, requiring different treatments. The authors of chapter 12 investigate the interactive factors that effect paper mill wastewater in order to optimize the coagulation process.

- The LCA approach, described in chapter 13, is useful for designing and constructing the most efficient wastewater treatment plants with the least environmental impact.

# LIST OF CONTRIBUTORS

**Iran Alemzadeh**
Chemical and petroleum engineering department, Sharif University of Technology, Azadi St, Tehran, Iran

**Rômulo A. Ando**
Departamento de Química Fundamental, Instituto de Química, Universidade de São Paulo, São Paulo, São Paulo, Brazil

**A. Asha**
Department of Chemical Engineering, A.C. Tech. Campus Anna University, Chennai, 600 025, India

**N. A. Atiqah**
Faculty of Agro Based Industry, Universiti Malaysia Kelantan,Jeli Campus, Locked Bag 100, 17600 Jeli, Kelantan, Malaysia

**Nader Bahramifar**
Department of Environmental Science, Faculty of Natural Resources, Tarbiat Modares University, P.O. Box 46414-356, Noor, Iran

**N. Balasubramanian**
Department of Chemical Engineering, A.C. Tech. Campus Anna University, Chennai, 600 025, India

**Francesca Beolchini**
Department of Life and Environmental Sciences, Polytechnic University of Marche, Ancona, Italy

**Noushin Birjandi**
Department of Environmental Science, Faculty of Natural Resources, Tarbiat Modares University, P.O. Box 46414-356, Noor, Iran

**Butenko E.**
Azov Sea State Technical University, Mariupol, Ukraine

**Nurdan Buyukkamaci**
Dokuz Eylul University, Engineering Faculty, Department of Environmental Engineering, Tinaztepe Campus, Buca, Izmir, Turkey

**Shri Chand**
Department of Chemical Engineering, Indian Institute of Technology Roorkee, Uttarakhand, India

**Hua-Wei Chen**
Department of Cosmetic Application & Management, St. Mary's Junior College of Medicine, Nursing and Management, Yilan 266, Taiwan

**Teng-Chien Chen**
Department of Chemical Engineering, National Cheng Kung University, Tainan 701, Taiwan; Sustainable Environment Research Center, National Cheng Kung University, Tainan 701, Taiwan

**Chyow-San Chiou**
Department of Environmental Engineering, National Ilan University, Ilan 260, Taiwan

**Kai-Jen Chuang**
Department of Public Health, School of Medicine, College of Medicine, Taipei Medical University, Taipei 110, Taiwan; School of Public Health, College of Public Health and Nutrition, Taipei Medical University, Taipei 110, Taiwan

**Benedito Corrêa**
Departamento de Microbiologia, Instituto de Ciências Biomédicas II, Universidade de São Paulo, São Paulo, São Paulo, Brazil

**Nuria Garcia-Mancha**
Sección Departamental de Ingenieria Quimica, Universidad Autonoma de Madrid, Francisco Tomas y Valiente 7, Madrid, 28049, Spain

**Ruo-Lin Huang**
Department of Chemical Engineering, National Cheng Kung University, Tainan 701, Taiwan

**Yao-Hui Huang**
Department of Chemical Engineering, National Cheng Kung University, Tainan 701, Taiwan; Sustainable Environment Research Center, National Cheng Kung University, Tainan 701, Taiwan

**A. Kapustin**
Azov Sea State Technical University, Mariupol, Ukraine

**Ya-Fen Lin**
Department of Environmental Engineering, National Ilan University, Ilan 260, Taiwan

**Eduardo Alberto López-Maldonado**
Centro de Investigación y Desarrollo Tecnológico en Electroquímica (CIDETEQ), Carretera Libre Tijuana-Tecate km. 26.5, Parque Industrial El Florido, 22500 Tijuana, BC, Mexico

**Chih-Ming Ma**
Department of Cosmetic Application & Management, St. Mary's Junior College of Medicine, Nursing and Management, Yilan 266, Taiwan

**A. Malyshev**
SASOL Germany GmbH

**V. Moganaragi**
Faculty of Agro Based Industry, Universiti Malaysia Kelantan,Jeli Campus, Locked Bag 100, 17600 Jeli, Kelantan, Malaysia

**Angel F. Mohedano**
Sección Departamental de Ingenieria Quimica, Universidad Autonoma de Madrid, Francisco Tomas y Valiente 7, Madrid, 28049, Spain

**Prasenjit Mondal**
Department of Chemical Engineering, Indian Institute of Technology Roorkee, Uttarakhand, India

**Victor M. Monsalvo**
Sección Departamental de Ingenieria Quimica, Universidad Autonoma de Madrid, Francisco Tomas y Valiente 7, Madrid, 28049, Spain

**A. Muthukrishnaraj**
Department of Chemical Engineering, A.C. Tech. Campus Anna University, Chennai, 600 025, India

**Cláudio A. Oller do Nascimento**
Departamento de Engenharia Química, Politécnica, Universidade de São Paulo, São Paulo, São Paulo, Brazil

**Adrián Ochoa-Terán**
Centro de Graduados e Investigación en Química del Instituto Tecnológico de Tijuana, Boulevard Alberto Limón Padilla s/n, Mesa de Otay, 22500 Tijuana, BC, Mexico

**Mercedes Teresita Oropeza-Guzmán**
Centro de Investigación y Desarrollo Tecnológico en Electroquímica (CIDETEQ), Carretera Libre Tijuana-Tecate km. 26.5, Parque Industrial El Florido, 22500 Tijuana, BC, Mexico; Centro de Graduados e Investigación en Química del Instituto Tecnológico de Tijuana, Boulevard Alberto Limón Padilla s/n, Mesa de Otay, 22500 Tijuana, BC, Mexico

**Chiara Pennesi**
Department of Life and Environmental Sciences, Polytechnic University of Marche, Ancona, Italy

**Ricky Priambodo**
Department of Chemical Engineering, National Cheng Kung University, Tainan 701, Taiwan

**Daniel Puyol**
Sección Departamental de Ingenieria Quimica, Universidad Autonoma de Madrid, Francisco Tomas y Valiente 7, Madrid, 28049, Spain

**Hayfa Rajhi**
Departamento de Biologia Molecular, Universidad Autonoma de Madrid, Darwin 2, Madrid, 28049, Spain

**Juan J. Rodriguez**
Sección Departamental de Ingenieria Quimica, Universidad Autonoma de Madrid, Francisco Tomas y Valiente 7, Madrid, 28049, Spain

**Mahdieh Safa**
Chemical and petroleum engineering department, Sharif University of Technology, Azadi St, Tehran, Iran

**Marcia R. Salvadori**
Departamento de Microbiologia, Instituto de Ciências Biomédicas II, Universidade de São Paulo, São Paulo, São Paulo, Brazil

**Ravi Shankar**
Department of Chemical Engineering, Indian Institute of Technology Roorkee, Uttarakhand, India

**Lovjeet Singh**
Department of Chemical Engineering, Indian Institute of Technology Roorkee, Uttarakhand, India

**Keerthi Srinivas**
Department of Chemical Engineering, A.C. Tech. Campus Anna University, Chennai, 600 025, India

**Cecilia Totti**
Department of Life and Environmental Sciences, Polytechnic University of Marche, Ancona, Italy

**Manouchehr Vossoughi**
Chemical and petroleum engineering department, Sharif University of Technology, Azadi St, Tehran, Iran

**Y. C. Wong**
Faculty of Agro Based Industry, Universiti Malaysia Kelantan,Jeli Campus, Locked Bag 100, 17600 Jeli, Kelantan, Malaysia

**Habibollah Younesi**
Department of Environmental Science, Faculty of Natural Resources, Tarbiat Modares University, P.O. Box 46414-356, Noor, Iran

# INTRODUCTION

Water is essential to our planet's life. Protecting our water resources is a prerequisite for building a sustainable future. But we face significant challenges.

Water use is inextricably linked to energy use. Water plays an essential role in many, if not most, manufacturing facilities, since most industries are dependent on processes that ultimately produce wastewaters. In a world facing a water-scarcity crisis, much research and development is focusing on decreasing industries' water-use footprint.

This compendium volume looks briefly at several select industries—industries that produce petrochemicals, metal industries, the semiconductor industry, and paper and pulp industries—and investigates various water treatment processes for each. These include microbial biotechnologies, ozone-related processes, adsorption, and photochemical reactions, among others.

Obviously, this is by no means an exhaustive or comprehensive coverage of the topic; the scope of this compendium is not great enough to investigate all treatment methods across a wider range of industries. Instead, the research gathered here is intended to be a starting point for further investigation. It also offers specific directions to follow further, expanding the potential of these particular studies into a wider frame of research work.

*—Victor M. Monsalvo*

In chapter 1, the authors focus on the anaerobic biodegradation of wastewater from used industrial oils (UIO) recovery using a bench-scale expanded granular sludge bed reactor (EGSB) at room temperature. Biodegradability tests showed that this wastewater can be partially biodegraded under anaerobic conditions at mesophilic temperature. Low concentrations of wastewater caused an incremented specific activity of the acetoclastic and the hydrogenotrophic methanogens. Anaerobic biodegradation at room temperature is feasible at organic loading rates (OLR) lower than 5.5 g COD $L^{-1}$ $d^{-1}$. A further increase of the OLR to around 10 g COD $L^{-1}$ $d^{-1}$ had a detrimental effect on the system performance, making it necessary to work at mesophilic conditions. The authors conclude that anaerobic treatment using an EGSB reactor is a feasible option for treating UIO wastewater. Long-term treatment caused a specialization of the granular sludge, modifying substantially its microbial composition. Methane production was even stimulated by the addition of UIO wastewater at low concentrations.

Next, in chapter 2, the authors develop a method of synthesis of layered double hydroxides (LDH) of different composition. They investigate the processes of adsorption of phenols on LDH variable composition. Finally, they design kinetic parameters for the processes of phenol adsorption.

Chapter 3 studies a novel implementation of a hybrid membrane bioreactor (HMBR), utilized as combination of rotating biological contractor (RBC) and an external membrane, as a new biological system for oily wastewater treatment. The authors evaluate chemical oxygen demand (COD) and total petroleum hydrocarbon (TPH) as factors of biodegradability. Both factors are compared with each other for different hydraulic retention times (HRTs) and petroleum pollution concentrations in RBC and HMBR. The ratio of TPH to COD of molasses was varied between 0.2 to 0.8 at two HRTs of 18 and 24 hours, while the temperature, pH, and dissolved oxygen were kept in the range of 20–25°C, 6.5–7.5, and 2–3.5 mg/l, respectively. The best TPH removal efficiency (99%) was observed in TPH/COD = 0.6 and HRT = 24 hr in HMBR, and removal efficiency was decreased in the ratios above 0.6 in both bioreactors. The experimental results showed that HMBR had higher treatment efficiency than RBC at all ratios and HRTs.

The use of dried and re-hydrated biomass of the seagrass *Posidonia oceanica* was investigated in chapter 4 as an alternative and low-cost biomaterial for removal of vanadium(III) and molybdenum(V) from wastewaters. Initial characterization of this biomaterial identified carboxylic groups on the cuticle as potentially responsible for cation sorption, and confirmed the toxic-metal bioaccumulation. The combined effects on biosorption performance of equilibrium pH and metal concentrations were studied in an ideal single-metal system and in more real-life multicomponent systems. There were either with one metal (vanadium or molybdenum) and sodium nitrate, as representative of high ionic strength systems, or with the two metals (vanadium and molybdenum). For the single-metal solutions, the optimum was at pH 3, where a significant proportion of vanadium was removed (ca. 70%) while there was ca. 40% adsorption of molybdenum. The data obtained from the more real-life multicomponent systems showed that biosorption of one metal was improved both by the presence of the other metal and by high ionic strength, suggesting a synergistic effect on biosorption rather than competition. The authors then used the data to develop a simple multi-metal equilibrium model based on the noncompetitive Langmuir approach, which was successfully fitted to experimental data and represents a useful support tool for the prediction of biosorption performance in such real-life systems. Overall, the results suggest that biomass of *P. oceanica* can be used as an efficient biosorbent for removal of vanadium(III) and molybdenum(V) from aqueous solutions. This process thus offers an eco-compatible solution for the reuse of the waste material of leaves that accumulate on the beach due to both human activities and to storms at sea.

The electroplating copper industry in Taiwan discharges huge amounts of wastewater, causing serious environmental and health damage. Chapter 5 applies to this situation research into the electrical copper recovery system. The authors study the electrotreatment of an industrial copper wastewater ($[Cu] = 30000 \, mg \, L^{-1}$) with titanium net coated with a thin layer of $RuO2/IrO2$ (DSA) reactor. The optimal result for simulated copper solution was 99.9% copper recovery efficiency in current density $0.585 \, A/dm2$ and no iron ion. Due to high concentrations of iron and chloride ions in real industrial wastewater, the copper recovery efficiency went down to 60%. Although, the copper recovery efficiency was not as high as in the simulated copper solution, high environmental economic value was included in the technology. The authors conclude that pretreating the wastewater with iron is a necessary step, before the electrical recovery copper system.

In the study described in chapter 6, a natural process was developed using a biological system for the biosynthesis of nanoparticles (NPs) and the yeast *Rhodotorula mucilaginosa* for the removal of copper from wastewater by dead biomass. Dead and live biomass of *Rhodotorula mucilaginosa* was used to analyze the equilibrium and kinetics of copper biosorption by the yeast in function of the initial metal concentration, contact time, pH, temperature, agitation, and inoculum volume. Dead biomass exhibited the highest biosorption capacity of copper, 26.2 mg g$^{-1}$, which was achieved within 60 min of contact, at pH 5.0, temperature of 30°C, and agitation speed of 150 rpm. The equilibrium data were best described by the Langmuir isotherm, and kinetic analysis indicated a pseudo-second-order model. The authors determined the average size, morphology, and location of NPs biosynthesized by the yeast by scanning electron microscopy (SEM), energy dispersive X-ray spectroscopy (EDS), and transmission electron microscopy (TEM). The shape of the intracellularly synthesized NPs was mainly spherical, with an average size of 10.5 nm. The X-ray photoelectron spectroscopy (XPS) analysis of the copper NPs confirmed the formation of metallic copper. The authors conclude that the dead biomass of *Rhodotorula mucilaginosa* may be considered an efficient bioprocess, being fast and low-cost. It is also probably nano-adsorbent for the production of copper nanoparticles in wastewater in the bioremediation process.

Tetramethyl ammonium hydroxide (TMAH) is an anisotropic etchant used in the wet etching process of the semiconductor industry and is hard to degrade by biotreatments when it exists in wastewater. The study in chapter 7 evaluated the performance of a system combined with ultraviolet, magnetic catalyst ($SiO_2/Fe_3O_4$) and $O_3$, denoted as $UV/O_3$, to TMAH in an aqueous solution. The mineralization efficiency of TMAH under various conditions follows the sequence: $UV/O_3 > UV/H_2O_2/O_3 > H_2O_2/SiO_2/Fe_3O_4/O_3 > H_2O_2/O_3 > SiO_2/Fe_3O_4/O_3 > O_3 > UV/H_2O_2$. The results suggest that $UV/O_3$ process provides the best condition for the mineralization of TMAH (40 mg/L), resulting in 87.6% mineralization, at 60 min reaction time. Furthermore, the mineralization efficiency of $SiO_2/Fe_3O_4/H_2O_2/O_3$ was significantly higher than that of $O_3$, $H_2O_2/O_3$, and $UV/H_2O_2$. More than 90% of the magnetic catalyst was recovered and easily redispersed in a solution for reuse.

In the study described in chapter 8, the treated and untreated effluents samples from semiconductor industry were collected and their physical characteristics (such as temperature, pH, turbidity, total suspended solid (TSS), conductivity) and chemical characteristics (such as salinity, dissolved oxygen (DO), biological oxygen demand (BOD), chemical oxygen demand (COD), chlorine dioxide, and heavy metal) were analyzed. The results showed that the semiconductor industry fulfilled the standard of law that was established in order to protect the environment.

The authors of chapter 9 improved the efficiency of coagulation-flocculation process used for semiconductor wastewater treatment by selecting suitable conditions (pH, polyelectrolyte type, and concentration) through zeta potential measurements. Under this scenario, the zeta potential, $\zeta$, is the right parameter that allows studying and predicting the interactions at the molecular level between the contaminants in the wastewater and polyelectrolytes used for coagulation-flocculation. Additionally, this parameter is a key factor for assessing the efficiency of coagulation-flocculation processes based on the optimum dosages and windows for polyelectrolytes coagulation-flocculation effectiveness. In this chapter, strategic pH variations allowed the prediction of the dosage of polyelectrolyte on wastewater from real electroplating baths, including the isoelectric point (IEP) of the dispersions of water and commercial polyelectrolytes used in typical semiconductor industries. The results showed that there is a difference between polyelectrolyte demand required for the removal of suspended solids, turbidity, and organic matter from wastewater (23.4 mg/L and 67 mg/L, resp.). The authors also conclude that the dose of polyelectrolytes and coagulation-flocculation window to achieve compliance with national and international regulations (the EPA in United States and SEMARNAT in Mexico) is influenced by the physicochemical characteristics of the dispersions and treatment conditions (pH and polyelectrolyte dosing strategy).

In recent years, extensive work has been conducted to improve the biodegradability index (BI) of effluent to enhance the efficiency of biochemical treatment. Advanced oxidation processes are one among the various methods primer to improve the BI of organic effluent. In the investigation described in chapter 10, experiments such as electrocoagulation, electro-oxidation, and photochemical processes were carried out to treat pulp and paper wastewater in a wide range of operating conditions. The effect of individual parameters on BI improvement was critically examined. The authors noticed that electro-oxidation method yields maximum biodegradability improvement (0.13–0.42) within a process time of 35 minutes.

Chapter 11 deals with the removal of lignin (expressed as COD removal) from synthetic wastewater through electrocoagulation in a batch reactor, using aluminum as a sacrificial electrode. The authors investigate the effects of various parameters such as current density, pH, NaCl concentration, and treatment time on the removal of COD from synthetic wastewater to determine the most suitable process conditions for maximum removal of COD (lignin). They use a central composite design (CCD) to create the experimental conditions for developing mathematical models to correlate the removal efficiency with the process variables. The most suitable conditions for the removal of lignin from synthetic solution were found to be: a current density of $100A/m_2$; a pH of 7.6; NaCl concentrations of 0.75mg/l; and a treatment time of 75 minutes. The authors' proposed model gives prediction on COD (lignin) removal with the error limit around +9 to −7%

In chapter 12, a coagulation process was used to treat paper-recycling wastewater with alum, coupled with poly aluminum chloride (PACl) as coagulants. The authors investigated the effect of each four factors, viz. the dosages of alum and PACl, pH and chemical oxygen demand (COD), on the treatment efficiency. They describe the influence of these four parameters using response-surface methodology under central composite design, and they consider the efficiency of reducing turbidity, COD and the sludge volume index (SVI) as the responses. The optimum conditions for high treatment efficiency of paper-recycling wastewater under experimental conditions were reached with numerical optimization of coagulant doses and pH, with 1,550 mg/l alum and 1,314 mg/l PACl and 9.5, respectively, where the values for reduction of 80.02% in COD, 83.23% in turbidity, and 140 ml/g in SVI were obtained.

In the final chapter, the author uses the Life Cycle Assessment (LCA), a "cradle-to-grave" approach, to identify energy use, material input, and waste generated, starting with the acquisition of raw materials and ending with the final disposal of a product from a particular facility. His review consists of four main activities: goal and scope, inventory analysis, impact assessment, and interpretation. In order to design and construct the most appropriate wastewater treatment plants, he points to the LCA approach as a useful tool for evaluating wastewater treatment techniques. He concludes that further, more detailed investigation is called for into the benefits and harms of each application in order to determine alternative, more environmentally friendly treatment methods using techniques that consume the least amount of energy.

# PART I

# INDUSTRIAL PETROCHEMICALS

Refinery and petrochemical plants generate solid waste and sludge composed of organic, inorganic compounds, including heavy metals. These wastewaters may contain polycyclic and aromatic hydrocarbons, phenols, metal derivatives, surface active substances, sulphides, naphthylenic acids and other chemicals. With ineffective purification systems, these toxic products accumulate in the receiving water, with serious consequences to the ecosystem.

Other industries besides refineries also produce petrochemical wastewater. Whether from food industries or the print industry, electronic industries or the textile industry, petrochemicals find their way into the world's water via wastewater streams. As a result sustainable treatment solutions for wastewater have become more and more important.

# CHAPTER 1

# Anaerobic Treatment of Wastewater from Used Industrial Oil Recovery

NURIA GARCIA-MANCHA, DANIEL PUYOL,
VICTOR M. MONSALVO, HAYFA RAJHI,
ANGEL F. MOHEDANOA AND JUAN J. RODRIGUEZ

## 1.1 INTRODUCTION

Huge quantities of industrial oil are consumed as a result of their extensive application (more than 310 kt consumed in Spain in 2009) in a broad range of industrial processes. Once used, the resulting oils are considered toxic and hazardous showing fairly low chemical or biological degradability. The amount of these residues has been estimated at around 90 kt in Spain (2009). The used industrial oils (UIO) contain hazardous and toxic compounds, e.g. naphthalene, benzene derivatives and toluene, whose presence in the environment can cause severe damage. The waste hierarchy postulated in the 2008/98/EC European Directive includes valorization as one preferential strategy for waste management. [1] In Spain, recycling of 65% of UIO has been postulated as an objective. [2] Although it is still early to assess the impact of the new regulations, it is expected that the legal framework will reinforce oil recycling and regeneration objectives that will result in additional volumes of aqueous off-streams from these

*Anaerobic Treatment of Wastewater from Used Industrial Oil Recovery.* © Garcia-Mancha, N., Puyol, D., Monsalvo, V. M., Rajhi, H., Mohedano, A. F. and Rodriguez, J. J. (2012), J. Chem. Technol. Biotechnol., 87. 1320–1328. doi. 10.1002/jctb.3753. Used with the authors' permission.

operations, usually including different stages of metal precipitation, extraction and distillation. [3] These effluents are characterized by high contents of organic solvents, hydrocarbons and suspended solids (SS), relatively high viscosity and poor biodegradability, making their treatment by conventional biological systems difficult. Nevertheless, biological techniques developed in the last two decades have shown their potential to deal with a broad range of industrial wastewater. In particular, anaerobic systems are an attractive option for the treatment of high-strength wastewater with an associated potential for methane generation. [4] A first approach to the anaerobic treatment by a two-step process of a synthetic wastewater simulating the effluents from UIO recycling operations showed COD removal efficiencies higher than 83%. [5] However, application to real wastewater has not been reported so far.

Among the anaerobic systems, the upflow anaerobic sludge blanket (UASB) reactor is the most widely applied for industrial wastewater. However, the so-called expanded granular sludge bed (EGSB) reactor is a promising alternative in which the height to diameter ratio and the external recirculation rate are increased, improving the mixing and contact between wastewater and biomass. [6] The viability of this system has been previously demonstrated in full-scale applications treating real wastewaters. [7] The height to diameter ratio plays an important role in the operation of EGSB reactors, and usually varies between 7 and 90 depending on the characteristics of the wastewater and the dimensions of the reactor. [8,9] Thus, EGSB systems have been reported to be adequate for dealing with hardly biodegradable wastewater, and to dampen the inhibition of the microbial activity caused by the presence of hazardous pollutants. These inhibition phenomena vary widely depending on the anaerobic inocula, wastewater composition, and working conditions. Some compounds commonly found in off-streams from UIO recovery, such as ammonia, sulphide, light metal ions, heavy metals, and organics, including alkil benzenes, phenol, alkil phenols, alkanes and alcohols have been reported to be inhibitors of anaerobic digestion. [10–13]

The identification of those microorganisms present in bacterial communities capable of degrading the organics present in this type of wastewater is of great interest for future industrial applications. Molecular biology techniques are useful to gain insights to the phylogenetic characterization of the microorganisms capable of removing chemicals from industrial wastewater. Some molecular techniques, like cloning and sequencing of extracted DNA and RNA, can provide exhaustive microbiological information. Among others, denaturing gradient gel electrophoresis (DGGE) is becoming routinely applied to microbial ecological studies to follow changes in the microbial population during the operation of anaerobic reactors. [14]

The aim of this work is to assess the anaerobic biological treatment of wastewater resulting from the regeneration process of UIO, determining the optimal operating conditions of an EGSB reactor. Biodegradability and the toxic effect of this wastewater on the methanogenic performance of the granular sludge under anaerobic conditions have also been evaluated. In addition, DGGE and sequencing of particular bands have been used to evaluate the evolution of the microbial population during the long-term experiment.

## 1.2 EXPERIMENTAL

### 1.2.1 BIOMASS SOURCE

Anaerobic granular biomass was collected from a full-scale UASB reactor treating sugar-beet wastewater (Valladolid, Spain). The granules had an average diameter of 0.5 mm and a specific methanogenic activity (SMA) of 0.46 g $CH_4$-COD $g^{-1}$ VSS $d^{-1}$.

### 1.2.2 WASTEWATER COMPOSITION

Wastewater was collected from a UIO recovery plant (Madrid, Spain). Themain characteristics of theUIO wastewater (10 samples tested)were: 29±5 gBOD5 $L^{-1}$, 107±47 gtotalCODL$^{-1}$, 94±38 g soluble COD $L^{-1}$, 2.87±1.07 g TSS $L^{-1}$, 2.67±1.12 g VSS $L^{-1}$ and pH10.5±0.8. Ethyleneglycol was identified as the most abundant chemical with a concentration of around 27.6 ± 0.9 g $L^{-1}$.

### 1.2.3 SMA DETERMINATION

Specific methanogenic activity (SMA) was measured using the Automatic Methane Potential Test System (AMPTS) developed by Bioprocess Control AB (Lund, Sweden). The AMPTS follows the same principles as the conventional methane potential test, thus making the results comparable with standard methods. Methane released from the digestion bottles is analyzed using a wet gas-flow measuring system with a multi-flow cell arrangement. This measuring device works according to the principle of liquid displacement and can monitor an ultra-low gas flow, where a digital pulse is generated when a defined volume of gas flows through the system. It only registers methane flow, since other gas components, such as $CO_2$ and $H_2S$, are removed by an alkaline solution. A data acquisition system is incorporated. [15] SMA values were calculated by the Roediger model [16] according to a previous work. [6]

### 1.2.4 BIODEGRADABILITY TESTS AND METHANOGENESIS STIMULATION EXPERIMENTS

Biodegradability tests were performed for 32 d using a nonadapted sludge by adding wastewater diluted at different ratios ranging from 6.25 to 50% v/v, which corresponds to COD values from 4 to 32 g $L^{-1}$. Stimulation of acetoclastic and hydrogenotrophic methanogenesis was carried out by activating the anaerobic sludge with acetate (4 g $CH_3COONa$ $L^{-1}$) or formiate (2 g $HCOONa$ $L^{-1}$), respectively, added to a standard methanogenic medium containing the following macronutrients (mg $L^{-1}$): $NH_4Cl_2$ (280), $K_2HPO_4$ (250), $KH_2PO_4$ (328), $MgSO_4 \cdot 2H_2O$ (100), $CaCl_2 \cdot 2H_2O$ (10) and yeast extract (4). This medium was supplemented with 5 mL $L^{-1}$ of a trace elements solution, reaching (μg $L^{-1}$): $FeCl_2 \cdot 4H_2O$ (2000), $H_3BO_3$ (50), $ZnCl_2$ (50), $CuCl_2 \cdot 2H_2O$ (38), $MnCl_2 \cdot 4H_2O$ (500), $(NH_4)_6Mo_7O_{24} \cdot 4H_2O$ (50), $AlCl_3 \cdot 6H_2O$ (90), $CoCl_2 \cdot 6H_2O$ (2000), $NiCl_2 \cdot 6H_2O$ (92), $Na_2SeO \cdot 5H_2O$ (162), EDTA (1000), resarzurin (200), $H_2SO_4$ 36% (1 μL $L^{-1}$). Buffer and alkalinity source was incorporated by adding 1 g $NaHCO_3$ $g^{-1}$ COD. The UIO wastewater was added to both media at COD concentrations ranging from 0.125 to 2 g $L^{-1}$. The biodegradability tests andmethanogenic stimulation experiments were performed at $30 \pm 1$ °C in duplicate, using the AMPTS.

The contribution of adsorption was evaluated in biomass samples after extraction with Soxhelt following the US-EPA 8041 method. Tests of volatilization were performed under identical operating conditions to those in the biodegradation experiments but in the absence of biomass. The results reported are the average values from duplicate runs, the standard errors being always lower than 10%.

## 1.2.5 EXPERIMENTAL SETUP FOR LONG-TERM EXPERIMENT

Experiments in continuous mode were carried out using a 5.2 L EGSB reactor with an internal diameter to height ratio of 1 : 7.2. The reactor was equipped with a gas–liquid–solid separator installed 15 cm below the exit. Wastewater was continuously fed, entering at the bottom of the reactor with recirculation, and the effluent was withdrawn from the top. $CO_2$ was removed from biogas using a Mariotte flask trap with 4 mol $L^{-1}$ NaOH solution, and $CH_4$ was measured with a wet gasometer (Schlumberger, Germany). [17] The reactor was operated at an upward flow rate of 2.5 m $h^{-1}$ and room temperature (17–21 °C) for 80 days. The EGSB reactor was inoculated with 100 g VSS $L^{-1}$ of granular sludge previously activated with a standard methanogenic medium for 30 d until high-activity stable performance was achieved. Macro and micronutrients, as well as $NaHCO_3$ (buffer and alkalinity source) to neutralize the influent were supplemented as detailed above. Organic loading rate (OLR) was varied between 0.5 and 10.5 g COD $L^{-1}$ $d^{-1}$ during the course of the experiment.

## 1.2.6 DNA EXTRACTION, PCR AND DENATURING GRADIENT GEL ELECTROPHORESIS (DGGE)

Granular sludge was resuspended in PBS, and cells were disrupted using a BIO101-Savant FP120 cell disrupter (Q BIOgene, Carlsbad, CA, USA) (six times for 40 s, each at 5.5 cycles $s^{-1}$). DNA was extracted using the FastDNA kit for soil (Q-BIOgene, Carlsbad, CA, USA). A fragment of the 16S rRNA gene was amplified by PCR with primer pairs 341(GC)-907R for Bacteria and 622(GC)-1492R for Archaea at annealing temperatures of 52 and 42 ∘C, respectively. [18] The amplification reaction was performed according to the taq DNA polymerase protocol (Promega, Madison, WIS, USA). The PCR conditions were as follows: 94 ∘C for 10 min; 30 cycles at 94 ∘C for 1 min, 72 ∘C for 3 min; and 72 ∘C for 10 min. The PCR products were analyzed using a D-Code Universal system (Bio-Rad, Hercules, CA, USA). An acrylamide solution with 6% p/v of porosity was used to cast a gel with denaturing gradients of urea/formamide ranging from 30 to 60% (100% = 7 mol $L^{-1}$ urea/40% v/v formamide). Electrophoresis was conducted in 1× TAE buffer solution at 200 V and 60 ∘C for 5 h. Bands detected by fluorescence using a UV transilluminator were excised and reamplified for sequencing. The sequences were automatically analyzed on an ABI model 377 sequencer (Applied Biosystems, Carlsbad, CA, USA) and were thereafter corrected manually. The sequences were compared with those listed in the Gen-Bank nucleotide sequence databases using Chromas 2.0 software. The BLAST search option of the National Center for Biotechnology Information (NCBI) (http://www.ncbi.nlm.nih.gov) was used to search for close evolutionary relatives in the GenBank database. Determination of the taxonomical hierarchy was performed using the Classifier tool from the Ribosomal Database Project (RDP) web page (http://rdp.cme.msu.edu/index.jsp) for the entire DNA sequences.

## 1.2.7 ANALYTICAL METHODS

Analyses of chemical oxygen demand (COD) and total and volatile suspended solids (TSS and VSS) were performed according to the APHA Standard Methods. [19] The identification of species in the influent and treated effluents was performed by gas chromatography/ion trapmass spectrometry (GC/MS, CP-3800/Saturn 2200, Varian, Santa Clara, CA, USA) with an autosampler injector (CP-8200, Varian, Agilent Technologies, Santa Clara, CA, USA) and solid-phase microextraction (Carbowax/Divinylbenzene, Yellow- Green). [17] Ethlylene-glycol, propylene-glycol, ethanol and volatile fatty acids (VFA) were quantified by HPLC coupled with a refraction index (HPLC/RI) detector (Varian, Agilent Technologies, Santa Clara, CA, USA) using sulfonated polystyrene resin in the protonated form (67H type) as the stationary phase (Varian Metacarb 67H 300–6.5 mm). [17]

## 1.3 RESULTS AND DISCUSSION

### 1.3.1 ANAEROBIC BIODEGRADABILITY

Figure 1(a) shows the time-evolution of COD and the cumulative methane production during the biodegradability tests.The specific COD consumption rate decreased from 0.34 to 0.11 g COD $g^{-1}$ VSS $d^{-1}$ as the COD of the fed stream was increased from 4 to 32 g COD $L^{-1}$. Simultaneously, the specific methane production rate decreased from 0.23 g CH4-COD $g^{-1}$ VSS $d^{-1}$ to almost zero under those same conditions. These results suggest the occurrence of inhibitory phenomena, since the activity values decreased when increasing the organic load of the UIO wastewater. Effects of adsorption or volatilization were negligible, so the COD removal can be attributed exclusively to biological degradation. A methane production of 1.15 g $CH_4$-COD was obtained at an UIO wastewater COD of 4 g $L^{-1}$, reaching a methanogenic potential of around 65%. In all cases, the methane production was lower than the corresponding theoretical value which supports the occurrence of inhibition. To learn more about the inhibition phenomena, the time-evolution of the main metabolites was analyzed. The time-evolution of ethylene-glycol and acetate is plotted in Fig. 1(b). As can be seen, ethylene-glycol was oxidized to acetate before methanization, which is in agreement with previous works. [20,21] Increasing the starting COD caused the accumulation of acetate in the medium, which can be explained by acetoclastic methanogenesis inhibition. The anaerobic oxidation of ethylene-glycol was negligible at 32 g $L^{-1}$ starting COD, confirming that at this COD the activity of the granular sludge was completely inhibited.

## 1.3.2 METHANOGENESIS STIMULATION

Methane production values from the methanogenesis stimulation experiments were used to calculate the specific methane production rates for acetoclastic and hydrogenotrophic methanogenesis shown in Fig, 2. As can be seen, the rate profiles of both processes differ considerably. A lag-time is observed in the acetoclastic experiments and the curves show a Gaussianlike profile with a maximum. In contrast, the maximum hydrogenotrophic methanogenesis rates occurred at the beginning of the experiments and a continuous quasi-linear decay takes place.

A detailed study of the stimulation of methanogenesis was carried out using the results of Fig. 2. Figure 3(a) shows the overall methane production in both the acetoclastic and hydrogenotrophic methanogenesis experiments. The methane production obtained when adding UIO at 2 g COD $L^{-1}$ was 1.2 and 3 times higher than that obtained in the blank experiments with acetate and formiate, respectively. This fact indicates that hydrogenotrophic methanogenesis stimulation produces an improvement of the UIO wastewater methanization. Acetoclastic and hydrogenotrophic SMA values followed different patterns with increasing UIO wastewater COD (Fig. 3(b)). The enhancement of the SMA upon stimulation of the hydrogenotrophicmethanogens was higher than that o served for the acetoclastic methanogens. Maximum values of 0.58 and 0.48 g $CH_4$-COD $g^{-1}$ VSS $d^{-1}$ were obtained for UIO wastewater COD of 0.5 and 2 g $L^{-1}$, respectively.

The complexity of the results from the acetoclastic experiments required a more detailed analysis. Maximum, average and initial specific acetoclastic rates were calculated and are depicted in Fig. 3(c). The UIO wastewater COD did not affect significantly the maximum specific rates, thus non-competitive inhibition phenomena in the acetoclastic methanogenesis seem negligible at the organic loads tested. However, the initial and average specific methanogenic rates increased, which could be related with the occurrence of the initial lag phase. Nevertheless, increasing UIO wastewater COD shortened the time required to reach the maximum specific rate (Fig. 3(d)), which suggests that the UIO wastewater has a beneficial effect on the adaptation of the granular sludge to produce methane from acetate.

### 1.3.3 EGSB REACTOR PERFORMANCE

The performance of the EGSB during long-term operation is depicted in Fig. 4. In order to avoid inhibition of the granular sludge, the OLR was gradually increased over 40 days at room temperature until reaching an OLR value of 10 g COD L$^{-1}$ d$^{-1}$. This operating strategy enabled COD removal efficiencies higher than 70% and methanogenic potential of around 0.55 g CH$_4$ g$^{-1}$ COD. Under these operating conditions, ethylene-glycol and the resulting acetate from its anaerobic oxidation were almost completely consumed. During the experiment, COD removal efficiencies suffered a slight reduction of 6 and 17% when applying OLR of 2.5 and 5.5 g COD L$^{-1}$ d$^{-1}$, respectively.

The increase of the OLR up to 10 g COD L$^{-1}$ d$^{-1}$ caused a drop in COD removal and methanogenic efficiency. Meanwhile, acetate concentrations close to 3 g L$^{-1}$ were detected in the resulting effluents. This indicates inefficient removal of the acetate generated from anaerobic oxidation, which can be related with inhibition of the acetoclastic methanogenesis caused by increased toxicity and the low operating temperature. To recover reactor performance, the temperature was set at 32 °C. A significant improvement in reactor performance was observed, resulting in COD removal and methanogenic efficiency of around 75 and 30%, respectively, while the acetate concentration was lowered below 0.5 g L$^{-1}$. From these results it can be concluded that controlling the temperature within the mesophilic range enabled further acclimation of the biomass, which was necessary to maintain reactor performance.

The composition of the UIO wastewater and the effluent from the EGSB reactor were analysed by GC/MS and HPLC/RI. Figure 5 depicts representative GC/MS chromatograms showing a fairly complex composition, summarized in Table 1. Most of the starting compounds were not detected in the resulting effluent after 80 days of continuous operation (Fig. 5(b)). However, some intermediate compounds were detected in the effluent, such as phenolic hydrocarbons as well as trace concentrations of acetate, propionate and ethanol, indicating incomplete anaerobic oxidation. The partial inhibition of biogas production and the decrease in biodegradability of the UIO wastewater can be caused by the presence of substituted phenolic compounds. [22, 23]

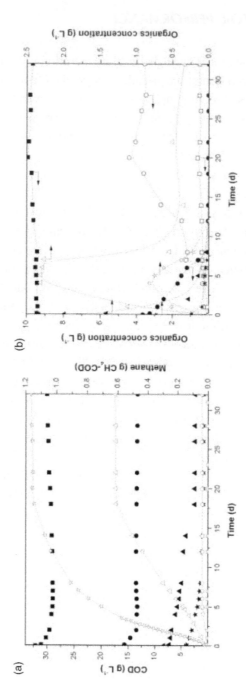

**FIGURE 1:** Time-evolution of (a) COD (solid symbols) and methane production (open symbols) and (b) ethylene-glycol (solid symbols) and acetate (open symbols) during the biodegradability assays. COD (g L$^{-1}$) of the fed wastewater: 32 (squares), 16 (circles), 8 (triangles) and 4 (stars).

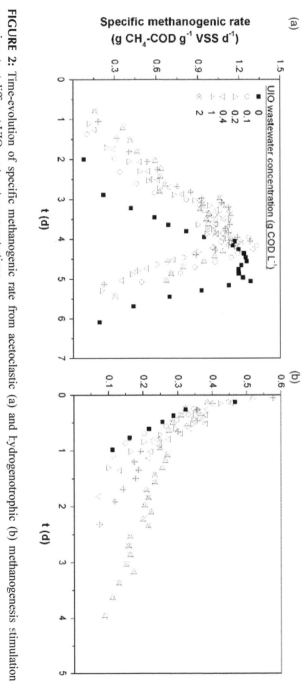

**FIGURE 2:** Time-evolution of specific methanogenic rate from acetoclastic (a) and hydrogenotrophic (b) methanogenesis stimulation experiments at different UIO wastewater concentrations.

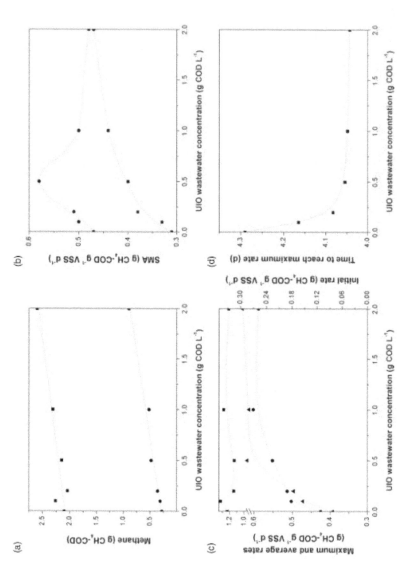

**FIGURE 3:** Overall methane production (a) and SMA (b) in the acetoclastic (■) and hydrogenotrophic (●) methanogenesis stimulation experiments. Maximum (■), average (●) and initial (▲) rates (c) and time to reach maximum rate (d) for acetoclastic methanogenesis at different UIO wastewater concentrations (COD).

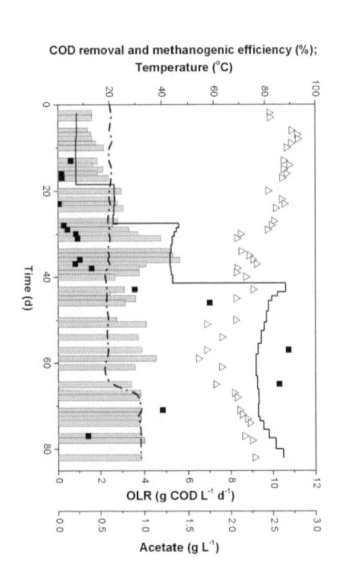

**FIGURE 4:** COD removal (△), methane production efficiency (bars), temperature (dash-dot line), OLR (solid line) and acetate concentration in the effluent (■) during the EGSB operation.

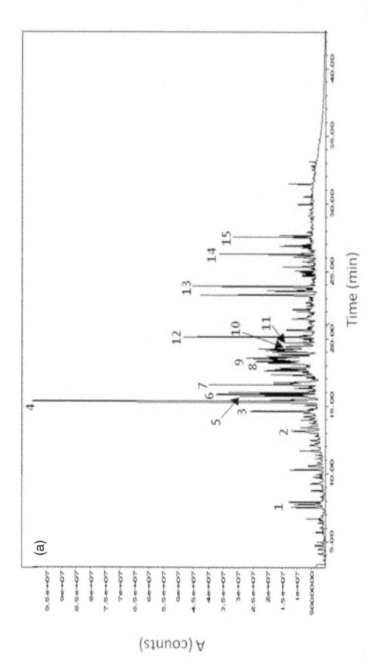

**FIGURE 5:** Representative GC/MS chromatograms of the UIO wastewater (32 g COD L$^{-1}$) (a) and the EGSB effluent (b).

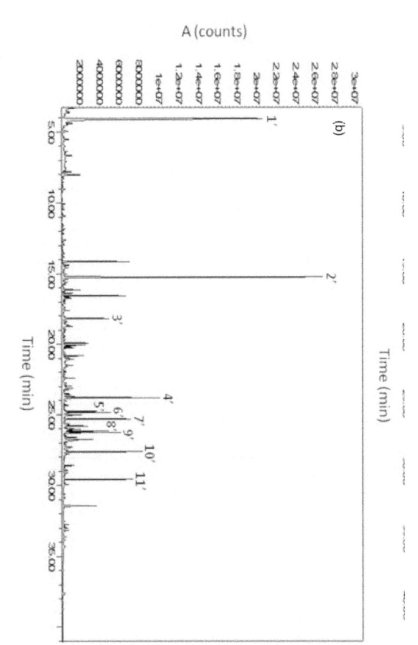

**FIGURE 5:** CONTINUED.

Evolution of the microbial population of the granular sludge Figure 6 shows the DGGE band patterns for the archaea and bacteria domains from the anaerobic granular sludge at the startup and after 90 days of EGSB reactor continuous operation. Type and number of bacteria and archaea band patterns changed significantly during the experiment because of the specialization of the granular sludge to treat UIO wastewater. Bands were excised, reamplified and sequenced for microbial identification by means of the NCBI and RDP databases (Table 2). A clear reduction of the archaeal diversity in the granular sludge took place and most of the identified *Methanosaeta sp.* (A2, A3, A4 and A5) species disappeared. However, the identified A1–A8 *Methanosaeta sp.* prevailed, since this species has been reported to be competitive in wastewater contaminated with hydrocarbons (NCBI access number HQ689197). New species of archaea (A6 and A7) belonging to the Thermoplasmatales order appeared during the experiment, which have been identified in UASB reactors treating alkanes-bearing wastewater [24] and usually appear in extremophilic environments. [25]

**TABLE 1:** Identification of peaks from Fig. 5.

| UIO wastewater | | Effluent | |
|---|---|---|---|
| Peak | Compound | Peak | Compound |
| 1 | 1-butanol | 1' | methyl isobutyl ketone |
| 2 | Tetradecane | 2' | 2-ethyl-1-hexanol |
| 3 | octyl-cyclopropane | 3' | acetophenone |
| 4 | 1-tetradecene | 4' | 2-methyl-phenol |
| 5 | Undecane | 5' | 2-ethyl-phenol |
| 6 | 3-methyl-2-pentene | 6' | 2,5-dimethyl-phenol |
| 7 | 1-nonanol | 7' | 2-(1-methylethyl)-phenol |
| 8 | 5-octadecene | 8' | 3,4-dimethyl-phenol |
| 9 | 2-methyl-2-undecanethiol | 9' | 4-ethyl-phenol |
| 10 | 1-bromo-pentadecane | 10' | p-tert-butyl-phenol |
| 11 | 1,4-dimethyl-cis-cyclooctane | 11' | indole |
| 12 | 3-tetradecene | HPLC/RI | Ethanol |
| 13 | Diphenyl ether | HPLC/RI | Acetate |
| 14 | 2-(1,1-dimethylethyl)-phenol | HPLC/RI | Propionate |
| 15 | 3,5-bis(1-methylethyl)-phenol | | |
| HPLC/RI | Ethlylene-glycol | | |
| HPLC/RI | Propylene-glycol | | |

**TABLE 2:** Identification of the DGGE bands for archaea (A) and bacteria (B) present in the sludge granules from the EGSB reactor.

| DGGE band | Sequence with higher homology* | Similarity (%) | NCBI GenBank access number | RDP taxonomical hierarchy | Ref. |
|---|---|---|---|---|---|
| A1 | Uncultured Methanosaeta sp. | 100% | HQ689197 | Methanosaeta sp. (100%) | U |
| A2 | Uncultured Methanosaeta sp. | 95% | JF754496.1 | Methanosaeta sp. (95%) | 29 |
| A3 | Uncultured Methanosaeta sp. | 97% | JF754496.1 | Methanosaeta sp (97%) | 29 |
| A4 | Uncultured Methanosaeta sp. | 96% | HQ689197 | Methanosaeta sp (90%) U | U |
| A5 | Methanosaeta conciii | 99% | CP002565.1 | Methanosaeta sp. (100%) | 30 |
| A6 | Uncultured Thermoplasmata archaeon | 99% | JF754533.1 | Archaea (100%), Thermoplasmatales (76%) 29 | 29 |
| A7 | Uncultured Thermoplasmata archaeon | 96% | JF754533.1 | Archaea (100%), Thermoplasmatales (50%) | 29 |
| A8 | Uncultured Methanosaeta sp. | 93% | HQ689197 | Archaea (100%, Methanosaeta sp. (80%) | U |
| B1 | Uncultured Bacterioidetes bacterium | 90% | HQ183944.1 | Bacteria (92%), Planctomycetes (16%) | 31 |
| B2 | NS | | | Bacteria (96%), Crysiogenetes (14%) | - |
| B3 | Uncultured Streptoococcus sp. | 96% | EU704223.1 | Bacteria (100%), Firmicutes (54%) | 26 |
| B4 | Uncultured Firmicutes bacterium | 95% | CU918169.1 | Bacteria (100%), Firmicutes (47%) | 27 |
| B5 | Uncultured bacterium | 96% | GU325923.1 | Bacteria (100%), Firmicutes (38%) | U |
| B6 | Uncultured Syntrophobacter sp. | 97% | EU888828.1 | Syntrophobacter sp. (97%) | 32 |
| B7 | Trichococcus flocculiformis | 98% | NR_042060.1 | Trichococcus sp. (100%) | 33 |
| B8 | NS | | | Bacteria (96%), Bacteroidetes (19%) | - |
| B9 | Uncultured bacterium | 95% | AB470353.1 | Bacteria(99%), Proteobacteria (75%), Gamma-proteobacteria (37%) | 34 |
| B10 | Uncultured bacterium | 95% | AB470353.1 | Bacteria(99%), Proteobacteria (63%), Gamma-proteobacteria (40%) | 34 |

**TABLE 2:** CONTINUED.

| | NS | | | |
|---|---|---|---|---|
| B11 | NS | | Bacteria (100%), Firmicutes (37%) | - |
| B12 | Uncultured bacterium | 92% | JF595815 | Bacteria (100%), Firmicutes (57%) | 28 |
| B13 | uncultured Chloroflexi bacterium | 100% | CU917962 | Bacteria (100%), Chloroflexi (33%) | 27 |
| B14 | Desulfomicrobium sp. | 99% | AY570692 | Desulfomicrobium sp. (100%) | 35 |
| B15 | Uncultured Aminanaerobia bacterium | 97% | CU917491 | Bacteria (100%), Synergistaceae (53%) | 27 |
| B16 | Uncultured bacterium | 93% | HQ453309 | Bacteria (100%), Synergistaceae (22%) | 36 |
| B17 | Aminiphilus circumscriptus | 99% | NR_043061 | Aminiphilus sp. (100%) | 37 |

\* NS=**BLAST** tool returns Non Significant responses for any query length.
U=**Unpublished.**

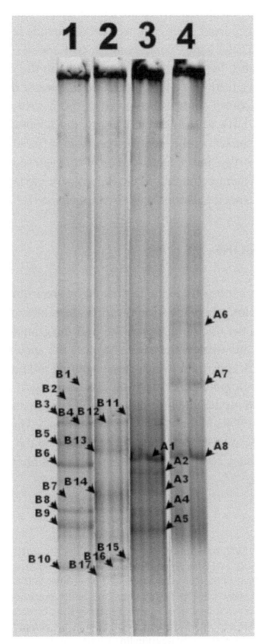

**FIGURE 6:** DGGE banding pattern for bacteria (1, 2) and archaea (3, 4) domains of the initial anaerobic granular sludge inoculum (1, 3) and during the long-term experiment (2, 4).

With regard to the bacteria domain, a clear specialization of the anaerobic consortium occurred since only two DGGE bands remained (B3–B11 and B4–B12). These bands correspond to two species belonging to the phylum Firmicutes, being one of them related with the genus *Streptococcus sp.* (B3–B11), [26] which have been found in waste digestion systems. [27,28] These species could be involved in the hydrolysis of the particulate matter of UIO wastewater. Owing to the adaptation of anaerobic granular sludge the rest of the identified bacteria suffered a clear modification. It is noteworthy that several species of nitrogen-consuming bacteria emerged,two of them belonging to the Synergistaceae family (B15, B16), and other to the species *Aminiphilus circumscriptus* (B17).

## 1.4 CONCLUSIONS

UIO wastewater canbe efficiently biotreated by anaerobic granular sludge in an EGSB reactor, where the specialization of the biomass leads to significant changes in the microbial composition of the granular sludge. At low concentrations, this wastewater even enhances the specific methanogenic activity of both acetate- and hydrogen-consuming methanogens, allowing high methanogenic potential. Anaerobic biodegradation at room temperature is feasible at moderate OLR values, but it is necessary to increase the temperature for OLRs higher than $5.5 \text{ g COD L}^{-1} \text{ d}^{-1}$. The observed inhibition of anaerobic microbial activity can be caused by the presence of some identified inhibitory compounds such as substituted phenols.

## REFERENCES

1.  EU, DIRECTIVE 2008/98/EC OF THE EUROPEAN PARLIAMENT AND OF THE COUNCIL of 19 November 2008 on waste and repealing certain Directives. Official Journal of the European Union (22 November 2008).
2.  Spain, REAL DECRETO 679/2006, de 2 de junio, por el que se regula la gesti ´on de los aceites industriales usados. (Spanish Royal Decree 679/2006, of June 2, by which the gestion of the used industrial oils is regulated), Boletin Oficial del Estado. (Spanish Official State Bulletin, 3 June 2006).

3.  Boughton B and Horvath A, Environmental assessment of use oil management methods. Environ Sci Technol 38:353–358 (2004).
4.  Chan YJ, Chong MF, Law CL and Hassell DG, A review on anaerobicaerobic treatment of industrial and municipal wastewater. Chem Eng J 155:1–18 (2009).
5.  Alimahmoodi M and Mulligan CN, Optimization of the anaerobic treatment of a waste stream from an enhanced oil recovery process. Biores Technol 102:690–696 (2011).
6.  Puyol D,Mohedano AF, Sanz JL andRodr´ıguez JJ,ComparisonofUASB and EGSB performance on the anaerobic biodegradation of 2,4- dichlorophenol. Chemosphere 76:1192–1198 (2009).
7.  Seghezzo L, Zeeman G, van Lier JB, Hamelers HVM and Lettinga G, A review: the anaerobic treatment of sewage in UASB and EGSB reactors. Biores Technol 65:175–190 (1998).
8.  Zhang Y, Yan L, Chi L, Long X, Mei Z and Zhang Z, Startup and operation of anaerobic EGSB reactor treating palm oil mill effluent. J Environ Sci 20:658–663 (2008).
9.  Fang C, Boe K and Angelidaki I, Biogas production from potato-juice, a by-product from potato-starch processing, in upflow anaerobic sludge blanket (UASB) and expanded granular sludge bed (EGSB) reactors. Biores Technol 102:5734–5741 (2011).
10. Chen Y, Cheng JJ and Creamer KS, Inhibition of anaerobic digestion process: a review. Biores Technol 99:4044–4064 (2008).
11. Calli B, Mertoglu B, Inanc B and Yenigun O, Effects of high free ammonia concentrations on the performances of anaerobic bioreactors. Process Biochem 40:1285–1292 (2005).
12. O'Flaherty V, Mahony T, O'Kennedy R and Colleran E, Effect of pH on growth kinetics and sulphide toxicity thresholds of a range of methanogenic, syntrophic and sulphate-reducing bacteria. Process Biochem 33:555–569 (1998).
13. Lin CY and Chen CC, Effect of heavy metals on the methanogenic UASB granule. Water Res 33:409–416 (1999).
14. D´ıaz EE, Stams AJM, Amils R and Sanz JL, Phenotypic properties and microbial diversity of methanogenic granules from a full-Scale upflow anaerobic sludge bed reactor treating brewery wastewater. Appl EnvironMicrobiol 72:4942–4949 (2006).
15. Jiu L, System setup for biological methane potential test. SwedenWO 2010/120230 A1 (2010).
16. Edeline J, Anaerobic reactors (digestors) (Reacteurs anaerobies (digesterus)), in Biological Depollution of Wastewater Theory and Technology, ed by Cebedoc, Liege, Belgium (1980).
17. Puyol D, Monsalvo VM, Mohedano AF, Sanz JL and Rodriguez JJ, Cosmetic wastewater treatment by upflow anaerobic sludge blanket reactor. J HazardMater 185:1059–1065 (2011).
18. Chan OC, Liu WT and Fang HH, Study of microbial community of brewery-treating granular sludge by denaturing gradient gel electrophoresis of 16S rRNA gene. Water Sci Technol 43:77–82 (2001).
19. APHA, AWWA and WPCF, Standard Methods for the Examination of Water and Wastewater American Public Health Association (APHA), Washington DC, USA (1992).

20. Staples CA, Williams JB, Craig GR and Roberts KM, Fate, effects and potential environmental risks of ethylene glycol: a review. Chemosphere 43:377–383 (2001).
21. Veltman S, Shoenberg T and Switzenbaum MS, Alcohol and acid formation during the anaerobic decomposition of propylene glycolundermethanogenicconditions. Biodegradation43:113–118 (1998).
22. Olguin-Lora P, Puig-Grajales L and Razo-Flores E, Inhibition of the acetoclastic methanogenic activity by phenol and alkyl phenols. Environ Technol 24:999–1006 (2003).
23. Hernandez JE and Edyvean RGJ, Inhibition of biogas production and biodegradability by substituted phenolic compounds in anaerobic sludge. J HazardMater 160:20–28 (2008).
24. Mbadinga SM, Wang L-Y, Zhou L, Liu J-F, Gu J-D and Mu PB-Z, Microbial communities involved in anaerobic degradation of alkanes. Int Biodeter Biodegr 65:1–13 (2011).
25. Brochier-Armanet C, Forterre P and Gribaldo S, Phylogeny and evolution of the Archaea : one hundred genomes later. Curr Opin Microbiol 14:274–281 (2011).
26. Scheithauer BK, Wos-Oxley ML, Ferslev B, Jablonowski H and Pieper DH, Characterization of the complex bacterial communities colonizing biliary stents reveals a host-dependent diversity. ISME J 3:797–807 (2009).
27. Riviere D, Desvignes V, Pelletier E, Chaussonnerie S, Guermazi S and Weissenbach J, et al, Towards the definition of a core of microorganisms involved in anaerobic digestion of sludge. ISME J 3:700–714 (2009).
28. Garcia SL, Jangid K, Whitman WB and Das KC, Transition of microbial communities during the adaption to anaerobic digestion of carrot waste. Biores Technol 102:7249–7256 (2011).
29. WangLY, GaoCX, MbadingaSM, ZhouL, LiuJF and GuJD, et al, Characterization ofanalkane-degradingmethanogenicenrichment culture from production water of an oil reservoir after 274 days of incubation. Int Biodeterior Biodeg 65:444–450 (2011).
30. Barber RD, Zhang L, Harnack M,Olson MV, Kaul R and Ingram-Smith C, et al, Complete genome sequence of Methanosaeta concilii, a specialist inaceticlasticmethanogenesis. JBacteriol 193:3668–3669 (2011).
31. Liu J, Wu W, Chen C, Sun F and Chen Y, Prokaryotic diversity, composition structure, and phylogenetic analysis of microbial communities in leachate sediment ecosystems. Appl Microbiol Biotechnol 91:1659–1675 (2011).
32. Worm P, Fermoso FG, Lens PNL and Plugge CM, Decreased activity of a propionate degrading community in a UASB reactor fed with synthetic medium without molybdenum, tungsten and selenium. Enzyme Microb Technol 45:139–145 (2009).
33. Liu JR, Tanner RS, Schumann P, Weiss N,McKenzie CA and Janssen PH, et al, Emended description of the genus Trichococcus, description of Trichococcus collinsii sp. nov., and reclassification of Lactosphaera pasteurii as Trichococcus pasteurii comb. nov. and of Ruminococcus palustris as Trichococcus palustris comb. nov. in the low-G+C grampositive bacteria. Int J Syst Evol Microbiol 52:1113–1126 (2002).
34. Iguchi A, Terada T, Narihiro T, Yamaguchi T, Kamagata Y and Sekiguchi Y, In situ detection and quantification of uncultured members of the phylum Nitrospirae abundant in methanogenic wastewater treatment systems. Microbes Environ 24:97–104 (2009).

35. Grabowski A, Nercessian O, Fayolle F, Blanchet D and Jeanthon C, Microbial diversity in production waters of a low-temperature biodegraded oil reservoir. FEMS Microbiol Ecol 54:427–443 (2005).

36. ZhangM YX, Zhang T, Chen J and Xue D,Molecular characterization of the bacterial composition in two waste silk refining systems. World JMicrobiol Biotechnol 27:2335–2341 (2011).

37. Diaz C, Baena S, Fardeau ML and Patel BK, Aminiphilus circumscriptus gen. nov., sp. nov., an anaerobic amino-acid-degrading bacterium from an upflow anaerobic sludge reactor. Int J Syst Evol Microbiol 57:1914–1918 (2007).

# CHAPTER 2

# LDHs as Adsorbents of Phenol and Their Environmental Applications

E. BUTENKO, A. MALYSHEV, AND A. KAPUSTIN

## 2.1 INTRODUCTION

At the present an increase of phenol and its derivatives concentration in the environment takes place, due to discharge it into the water from inadequately treated wastewater of different enterprises, especially coke and petrochemical enterprises [1]. Phenol can cause different diseases of organisms, including humans. Phenol is particularly dangerous because of its relatively good solubility in water.

To reduce the penetration of phenol into the environment may take place primarily due to the effective treatment of industrial wastewater, in which the phenol and its derivatives are found.

There are various methods of purification of industrial wastewater from the dissolved phenol. Liquid extraction is used for the purification of wastewater containing phenols [2]. The feasibility of wastewater treatment through extraction is determined by the concentration of organic

*LDHs as Adsorbents of Phenol and Their Environmental Applications* © *Butenko E., Malyshev A., and Kapustin A.* American Journal of Environmental Protection, *vol. 2, no. 1 (2014): 11-15. doi: 10.12691/env-2-1-3. Used with permission of Science and Education Publishing.*

impurities. Extraction of phenols from wastewater is economically inefficient process, because of use of costly ethers and esters for the extraction of phenols from wastewater.

The radiation treatment is used to remove phenol from wastewater. During the $\gamma$-radiation ($Co^{50}$, $Cs^{137}$) oxidation and polymerization of organic and inorganic substances (including not biodegradable and toxic compounds) takes place. The deposition of colloidal and suspended solids, disinfection and deodorization take place too. Radiation treatment, as a fast single-stage process gives the complex effect [3].

$Co^{60}$ $\gamma$-irradiation of aqueous solutions containing 10 mg/l of phenol in a flow system with a dose of 0,48 W/kg for 20 min completely decomposes it to water and carbon dioxide, the decomposition rate is 0.25 mg/(l min). However, through radiation treatment the removal of phenol and its derivatives up to the level of MPC is not achieved.

The final purification achieving MPC (0.1 g/l) is only under the use of adsorption technology possible. Adsorption methods are widely used for wastewater treatment from dissolved organic substances. The advantage of this method is its high efficiency, and the possibility of sewage directly from many toxic substances. The effectiveness of adsorption reaches 80-95%, depending on the nature of the adsorbent, the adsorption surface and its accessibility, on the chemical structure of substance and its state in solution [4].

Most of all, activated carbons [5] are used as a sorbent for the removal of phenols. However, their use is limited by their high cost. In addition, the sorption process on activated carbon are physical process, the sorbed ions are not associated with a matrix by chemical bonds, which makes the disposal process risky due to the possible of the reverse process of desorption. Also, carbon sorbents have disadvantages such as long-term establishment of sorption equilibrium and the low degree of sorption.

The most promising sorbents are the sorbents based on clays, synthetic and natural [6]. They are cheap, accessible and efficient, universal sorbents, and they have a high absorption capacity, resistance to environmental influences and can serve as excellent carriers for mounting on the surface of various compounds at their modification.

The aim of our work is the search for new anionic sorbents, the use of which would allow effective remove of phenols from wastewater.

For this aim LDHs with variable composition were investigated, and their physical, chemical, as well as adsorption properties were studied. Also we studied the kinetics of ion-exchange sorption of phenol on the LDHs of different composition, the structure of LDHs and its changes during the phenol sorption.

## 2.2 EXPERIMENTAL SECTION

Synthesis of layered double hydroxides was carried out by the method described in [7, 8]. Samples were kindly provided by SASOL Germany GmbH.

The obtained sorbents have the following characteristics (Table 1).

**TABLE 1:** The characteristics of layered double hydroxides.

| Mg/Mg+Al, mol/mol | 0.52 | 0.72 | 0.81 | 0.86 |
|---|---|---|---|---|
| d, Å | 3.038 | 3.036 | 3.045 | 3.058 |
| c, Å | 22.62 | 22.67 | 22.81 | 23.40 |
| Specific surface area, $m^2/g$ | 250 | 200 | 192 | 180 |
| Pore volume, ml/g | 0.5 | 0.2 | 0.2 | 0.2 |
| Acidity, meq/g | 0.41 | 0.32 | 0.21 | 0.06 |
| Basicity, meq/g | 0.73 | 0.54 | 0.63 | 0.85 |
| $E_{din.}$, meq/g | 0.075 | 0.081 | 0.041 | 0.036 |
| $E_{stat.}$, meq/g | 0.38 | 0.41 | 0.10 | 0.08 |

The resulting LDHs were investigated to determine the surface basicity. Determination was carried out by titration in the presence of Hammett indicators [9]. Analysis were carried out by the following procedure. In a glass beaker (20 ml), mounted on a magnetic stirrer was placed a sample of a layered double hydroxide, and then a glass was filled with benzene. Then a glass cylinder, divided by a porous partition with a standard sample on it was placed in the beaker. Hammett indicator was added into benzene, and the color change can be observed on the surface of a standard sample. The followed indicators were used: bromothymol blue (pKa =

7.2), 2-chloro-4-nitroaniline (pKa = 17.2), 4-chloroaniline (pKa = 26.5), 2,4,6-trinitroanilin (pKa = 12 and 2), 2,4-dinitroaniline (pK = 15.0), 4-nitroaniline (pKa = 18.4). Calcined MgO was used as a comparison standard.

Structure of LDHs was studied by X-ray diffraction (Figure 1). The X-ray diffraction experiments were performed using a SIEMENS D-500 diffractometer with Co Kα - radiation. Special computer programs were used for smoothing, background correction and decomposition of overlapped diffraction peaks.

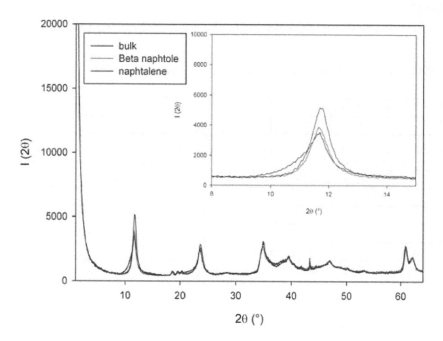

**FIGURE 1:** XRD patterns of LDH Mg/(Mg+Al) = 0,72 mol/mol (as-prepared, with naphthalene, with β-naphtol).

X-ray fluorescence spectroscopy (XRF), Shimadzu XRF-1700 sequential XRF spectrometer was used to determine the Mg/Al atomic ratios of the samples.

The nitrogen adsorption–desorption isotherms were recorded at 77 K on a Coulter SA 3100 automated gas adsorption system on samples previously degassed at 383 K for 7 h under vacuum. Specific surface areas (S-BET) were determined using the Brunauer–Emmett Teller (BET) method on the basis of adsorption data. The pore volume (Vp) values were determined by using the t-plot method of De Boer.

Thermal behavior was studied by thermo-analytical methods, the samples calcined on the air (the temperature was raised with 10 K/min to 1273 K).

The concentration of phenols was determined photometrically through the reaction of formation of the colored compound with 4-aminoantipyrine in an alkaline medium (pH=10) in the presence of ammonium persulfate. The concentration of alcohol, naphthalene and β-naphthol was determined by chromatography on a chromatograph HP 5890.

Sorption studies were performed in periodic conditions, loading the sorbent in a solution containing sorbed substances (alcohols, phenols, naphthalene derivatives), and samples were taken after vigorous mixing for some time.

## 2.3 RESULTS AND DISCUSSION

LDHs are products of isomorphic substitution of metal cations in the hydroxides of metal cations of higher oxidation degree [13], as shown in Figure 2.

The acid properties of phenol are beneficial for using of LDHs for the phenol adsorption is. Studying LDHs, unlike most natural clays, are not solid bases, not acids. The basic site in such compounds may be represented by a hydroxyl group, which is localized at the tetrahedral aluminum. Lewis basic sites are a lone electron pair of oxygen:

Bronsted site                    Lewis site

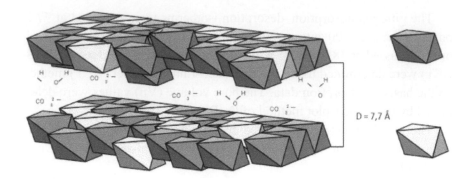

**FIGURE 2:** XRD patterns of LDH Mg/(Mg+Al) = 0,72 mol/mol (as-prepared, with naphthalene, with β-naphtol).

The presence of the basic sites of Brønsted and Lewis types allows the process of anion exchange in the inner space of layered double hydroxides. The anion-exchange reaction for phenol and its derivatives on layered double hydroxides proceeds as follows:

$$\equiv\!\!-OH + PhOH \longrightarrow \equiv\!\!-OPh + H_2O$$

Since phenol is an acid, so the represented reaction of anion exchange takes place almost irreversibly. In addition, phenol is a rather strong organic acid, so the process of ion-exchange takes place very quickly.

To determine the process parameters of sorption of phenols on LDH kinetic studies were carried out. Conducting of kinetic studies under the conditions far from the sorption equilibrium was predetermined by the fact that in real industrial conditions the sorption processes in exactly non-equilibrium conditions. The study was carried out in the mixed reactor

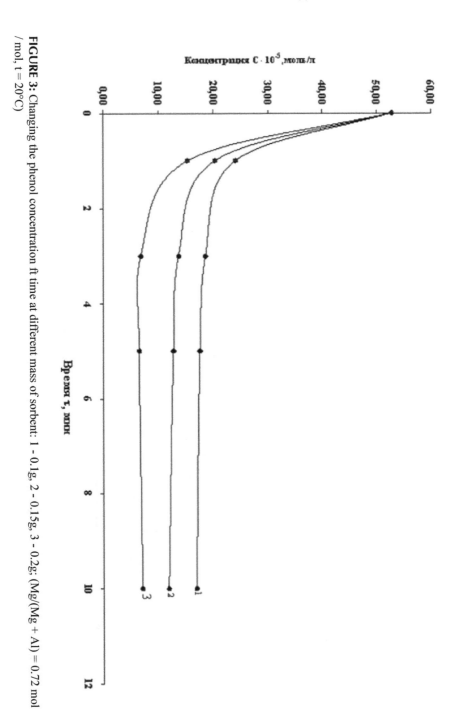

**FIGURE 3:** Changing the phenol concentration ft time at different mass of sorbent: 1 - 0.1g, 2 - 0.15g, 3 - 0.2g; (Mg/(Mg + Al) = 0.72 mol / mol, t = 20°C)

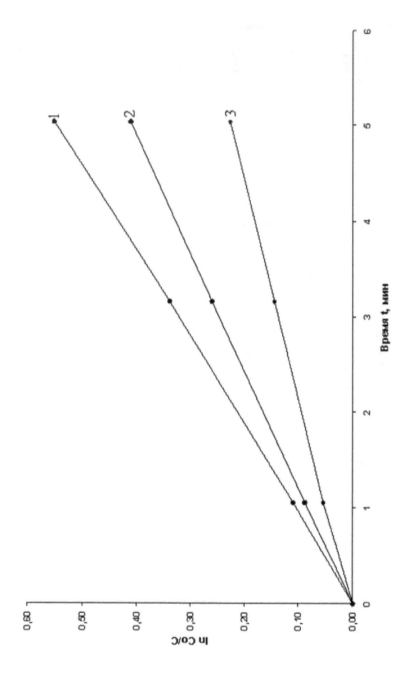

**FIGURE 4:** Dependence of $\ln(c_o/c)$ on the sorption time, 1 - 0.1g, 2 - 0.15g, 3 - 0.2g; (Mg/(Mg + Al)) = 0.72 mol/mol, t = 20°C)

**TABLE 2:** The dynamic capacity of LDHs (phenol).

| Mg/(Mg+Al), mol/mol | 0.52 | 0.72 | 0.81 | 0.86 |
|---|---|---|---|---|
| E, meq/g | 0.18 | 0.31 | 0.36 | 0.46 |

with periodic feeding, the concentration of phenol was determined by spectrophotometry. The obtained data are depicted on Figure 3.

At the initial moment, at high degree of conversion, the experimental results are well linearizing in the $\ln(c_0/c) - \tau$ coordinates (Figure 4), indicating the first order on adsorbate concentration.

Investigation of the kinetics of the reaction showed that the rate of sorption depends on the concentration of phenol and amount of LDHs. To replace the mass of the sorbent on the concentration of active sites (Cas) in a volume of solution used the values of the dynamic capacity of LDHs, defined in dynamic conditions (Table 2).

The kinetic equation of sorption of phenol is as follows:

$$\vartheta = k \cdot C_{PhOH} \cdot C_{as}$$

We investigated the adsorption capacity of phenol in reactions with sorbents with varying degrees of isomorphous substitution. Rate constants for phenol sorption for sorbents with different molar ratio of Mg / (Mg +

**TABLE 3:** Second order constants of sorption.

| Mg/(Mg+Al), mol/mol | 0.52 | 0.72 | 0.81 | 0.86 |
|---|---|---|---|---|
| $k \cdot 10^3$, l/mol $\cdot$ s | 0.83 | 1.29 | 1.67 | 1.70 |

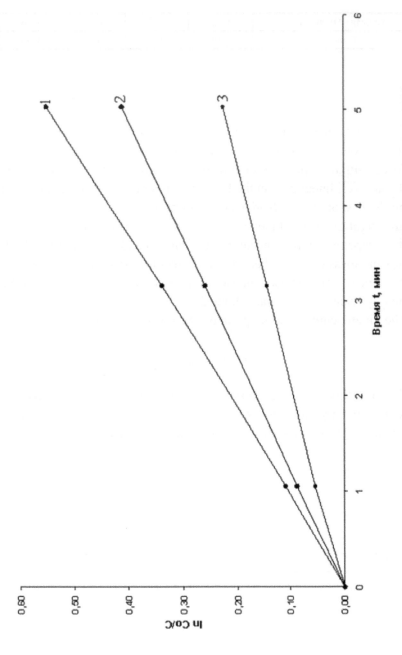

**Figure 5.** Changing of phenol concentration in time at different temperatures; 1 – 21°C, 2 – 40°C, 3 – 60°C, 4 – 75°C.

Al) have been calculated. The obtained values of the constants are presented in Table 3.

To determine the activation parameters of sorption of phenols on LDHs, the ion exchange at different temperatures has been studied. These data are presented as curves on Figure 5.

Since the reaction of a rather strong organic acid with a solid base proceeds quickly, the linearization performed at the initial moment of reaction. The rate constants of sorption of phenol at different temperatures were determined. The obtained data are presented in Table 4.

Dependence of the rate of phenol sorption on the temperature has been investigated only for the sorbent $Mg/(Mg+Al) = 0,72$ mol/mol, however, we can assume that for other sorbents will be observed a similar dependence.

The full kinetic equation for the phenol sorption of phenol sorbent has the form:

$$k = k_0 = 9.9 \cdot 10^3 \cdot e^{-29,1/RT}$$

The obtained value of activation energy suggests that the reaction proceeds in the diffusion region, but close enough to the kinetic region, due to the high acidity of phenol.

The reaction with phenol proceeds rapidly and after 15-20 minutes the dynamic equilibrium set. Due to the high acidity of phenol the equilibrium is strongly shifted toward the formation of products, the reaction proceeds to completion.

**TABLE 3:** Second order constants of sorption.

| T, K | 294 | 313 | 333 | 348 |
|------|-----|-----|-----|-----|
| k, $s^{-1}$ | 0.08 | 0.24 | 0.33 | 0.46 |
| E = 29.1 kJ/mol | | | $k_0 = 9.9 \cdot 10^3$ l/mol · s | |

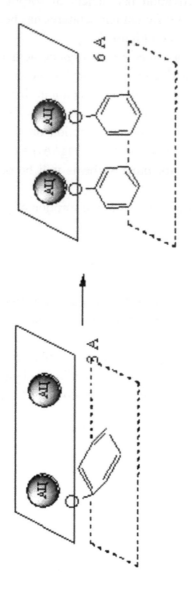

**FIGURE 6:** Change the location of the molecules of phenol in the LDHs).

However, if at the end of the sorption to leave the LDHs in a solution of phenol in a long time (48 hours), the degree of adsorption increases, a simultaneous increase the inter-planar distance from 3,03 to 5,76 Å. By our assumption, this is due to the reorientation of the aromatic rings in the inner space of LDHs (Figure 6), with increased availability of active sites.

## 2.4 CONCLUSIONS

1.  Physical-chemical properties of LDHs. It has shown that changing the composition of LDH the concentration and strength of active sites on their surface can be regulated, and thus we can regulate the sorption properties.
2.  The kinetics of phenol adsorption on the LDH of variable composition have been studied.
3.  The dependence of the kinetic parameters of the adsorption on the composition of LDHs has been shown.
4.  The kinetic and activation parameters of the sorption of phenol on the LDHs, which may be the basis for the calculation of process plants for the removal of phenols from industrial wastewater, have been determined.

## REFERENCES

1.  Rubio J. Overview of flotation as a wastewater treatment technique / J. Rubio, M. Souza, R. Smith //Minerals Engineering. 2002. Vol. 15, P. 139-155.
2.  Яковлев. С.В. Современные решения по очистке природных и сточных вод / Яковлев. С.В., Демидов О.В. // Экология и промышленность России. 1999. № 12. C. 12-15.
3.  Куркуленко С.С. Состояние обращения с отходами в Донецкой области / С.С. Куркуленко, Г.И. Бородай // Регион: проблемы и перспективы. 2002. Т. 8. С. 20-23.
4.  Roostaei N. Removal of phenol from aqueous solutions by adsorption. / N. Roostaei, F. Handan Tezel // Environmental Management, 2004. V. 70. P. 157-164.
5.  Arellano-Cárdenas S. Adsorption of Phenol and Dichlorophenols from Aqueous Solutions by Porous Clay Heterostructure (PCH). / S. Arellano-Cárdenas, T.

Gallardo-Velázquez, G. Osorio-Revilla, Ma. del Socorro López-Cortéz1 and B. Gómez-Perea1 // J. Mex. Chem. Soc., 2005.V. 49(3). P. 287-291.

6. Капустин А.Е. Катализ слоистыми двойными гидроксидами. / А.Е. Капустин // Научные проблемы современной технологии. 2007. № 16. С. 267-275.

7. Vaccari. A. Layered double hydroxides: present and future / A. Vaccari, V. Rives // Journal of Membrane Science. 2002.Vol. 9. P. 134-138.

8. Bolongini M. Mg/Al mixed oxides prepared by coprecipitation and sol-gel routes: a comparison of their physico-chemical features and performance in m-cresol methylation / M. Bolongini, F. Cavani, C. Perego // Microporous and Mesoporous Materials. 2003. V. 66. P 77-89.

9. Грекова Н.Н. Проблемы индикаторного титрования основных гетерогенных катализаторов / Н.Н. Грекова, О.В. Лебедева, А.Е. Капустин // Катализ и нефтехимия. 1996. № 2. С. 76-79.

10. Fagerlund G. Determination of specific surface by the BET method / G. Fagerlund // Materials and Structures. 2006. P 239-245.

11. Грег С.Н. Адсорбция, удельная поверхность, пористость / С.Н. Грег. М., 1990. 504 с.

12. Крайнов С.Р. Геохимические и экологические последствия изменения химического состава подземных вод под влиянием загрязняющих веществ / С.Р. Крайнов, Г.Ю. Фойгт, В.П. Закутин // Геохимия. 1991. №2. С.169-182.

13. Капустин А.Е. Неорганические аниониты / А.Е. Капустин// Успехи химии. 1991. Т. 60. № 12. С. 2685-2717.

# CHAPTER 3

# Biodegradability of Oily Wastewater Using Rotating Biological Contactor Combined with an External Membrane

MAHDIEH SAFA, IRAN ALEMZADEH,
AND MANOUCHEHR VOSSOUGHI

## 3.1 BACKGROUND

Nowadays, one of the major environmental problems is the oily wastewaters produced by industries, particularly by refineries. Disposal of oily wastewaters into the environment can result in environmental pollutions and serious damages to the ecosystem. Since conventional treatment processes are not sufficient to achieve the water quality requirements, advanced treatment processes are required [1].

The HMBR is an advanced technology which traditionally combines activated sludge as a suspended growth system with microfiltration (MF) or ultrafiltration (UF) membrane [2]. This process has now become an attractive choice for the treatment and reuse of industrial wastewaters such

*Biodegradability of Oily Wastewater Using Rotating Biological Contactor Combined with an External Membrane.* © *2014 Safa et al.; licensee BioMed Central Ltd.* Journal of Environmental Health Science and Engineering *2014, 12:117 doi:10.1186/s40201-014-0117-3. Creative Commons Attribution License (http://creativecommons.org/licenses/by/2.0).*

as paper mill; food production; fuel port [3]–[5] and municipal wastewaters [6],[7]. The HMBR process has been proved to have many advantages in comparison to conventional biological processes such as small footprint size of the treatment unit, reduced sludge production, complete retention of solids and flexibility of operation [8].

The initiative of the present research is substituting the suspended growth system with the attached growth system. Therefore, RBC (plus Kaldnes media) as an attached growth system was coupled with external UF membrane to treat oily wastewater.

The reason for choosing RBC can be related to many advantages of this reactor in treating wastewaters, particularly oily wastewaters, compared to the active sludge process. Among the advantages, one can include high efficiency of organic matter removal, resistance against organic and hydraulic shock loads and low energy consumption [9].

Experiments were carried out to compare the performance of the RBC and the HMBR in treating the oily wastewaters. After adjusting oil-eating microorganisms with system, the influence of some parameters as HRT, TPH and nutrients concentration on the performance of system were studied. The efficiency of two systems in removal of oily pollutants and organic matters produced by nutrients was also examined and compared.

## 3.2 METHODS

### 3.2.1 PHYSICAL PROPERTIES OF THE SYSTEM

Figure 1 shows an overview of the hybrid membrane bioreactor. HMBR is a combination of rotating biological contactor (plus Kaldnes media) and an external membrane. Effluent from bioreactor enters the membrane by a centrifugal pump and the sludge remained behind the membrane, which contains microorganisms, is returned to the bioreactor. Samples were collected from influent/effluent of RBC and effluent of the membrane. Physical properties of RBC and the membrane are shown in Tables 1 and 2, respectively.

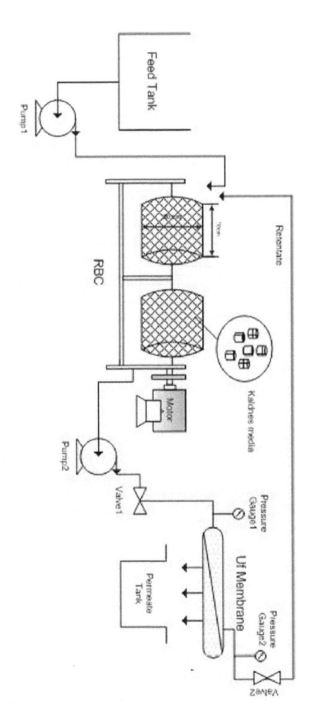

**FIGURE 1:** Overview of the hybrid membrane.

**TABLE 1:** Physical properties of the rotating biological contactor

| Length | 40 cm | Width | 27 cm |
|---|---|---|---|
| Height | 20 cm | Total surface | 2m² |
| Total volume | 21.6 lit | Effective volume | 18 lit |
| Cylinder diameter | 20 cm | Cylinder length | 10 cm |
| Stage | 2 | Rotational speed | 10 rpm |

**TABLE 2:** Physical properties of the membrane

| Membrane type | Ultra filtration |
|---|---|
| Membrane material | Polymer |
| Internal diameter | 1.24 cm |
| Effective length | 33 cm |
| Surface area | 0.0128 m² |

## 3.2.2 BIOREACTOR FEEDING

In this study, the sludge of second settling tank of the activated sludge process in Tehran refinery was used. At the first, RBC was prepared and 90% volume of its cylinders was filled with Kaldnes media. Then, some of the sludge plus some water was poured into the bioreactor so that the mixed liquor suspended solids (MLSS) in bioreactor became 1500 mg/l.

In order to grow and reproduce microorganisms and biofilm formation, the system was set up in batch process with COD = 1000 mg/l so it fed with carbon (molasses), nitrogen (urea) and phosphorus (ammonium phosphate) for 8 weeks. During the process, a combination of crude oil and gasoline with a ratio of 2/1 (petroleum pollutant in this state would have a wide range of hydrocarbons from C14 to C42) was added to the system for more adaption of microorganisms to the petroleum pollutant. Also, 5-15 μl/l of surfactant twin-80 was added to system so that bonds are formed between water and oil molecules. To speed up the microorganisms growth, some minerals were added to the sludge as well [10].

### 3.2.3 EXPERIMENTAL PROCESS

Once the biofilm with a thickness of about 4 mm was formed, the system was started up as continuous process at a HRT of 24 and 18 hours. The external membrane was connected to RBC. Subsequently, wastewater influent and effluent of RBC and effluent of HMBR were examined daily. These tests included the measurement of COD, MLSS, MLVSS, TPH, TSS, pH, temperature, and dissolved oxygen. All tests were performed according to the standard methods [11]. As time passes, membrane fouling causes such a permeate flux decline that the membrane needs to be refreshed. For this reason, bioreactor was switched off and the membrane was washed with water, NaOH 2%, and $HNO_3$ 1% [12].

### 3.3 RESULTS

### 3.3.1 EFFECT OF HYDRAULIC RETENTION TIME ON COD REMOVAL

Figure 2 shows the diagram of COD removal efficiency versus different concentrations of TPH at two HRTs (18 & 24 hr) for each reactor.

As shown in Figure 2 increasing the ratios of TPH/COD molasses has led to the reduction of COD removal efficiency in both reactors. This is resulted from the fact that increasing TPH/COD molasses makes microorganisms start to use oily hydrocarbons instead of nutrients produced by molasses.

On the other hand, in order to form bonds between water and oily pollutant molecules, some concentration of surfactant twin-80 was added to the system which helps the absorption of hydrocarbons by microorganisms.

As depicted in Figure 2 when the ratio of TPH/COD was greater than 0.6, the slope of the efficiency decrement was increased. This is due to inhibition caused by aromatics and hydrocarbons in oily wastewater. Furthermore, COD removal efficiency was increased for higher HRT. This is caused by the contact between nutrients and microorganisms for a longer retention time.

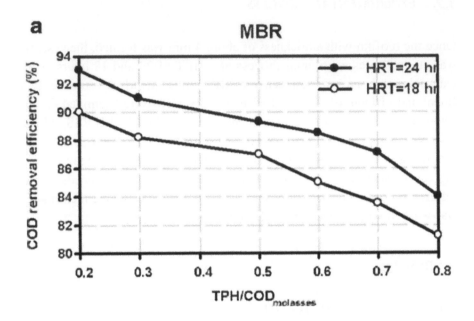

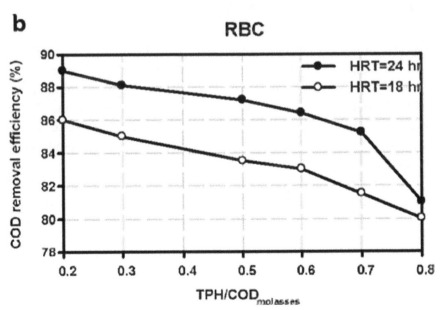

**FIGURE 2:** COD removal efficiency at two HRTs of 18 and 24 hours versus ratios of TPH/COD molasses in (a) MBR and (b) RBC.

## 3.3.2 THE EFFECT OF HRT ON TPH REMOVAL

TPH removal efficiency versus different concentrations of TPH at two HRTs (18 & 24 hr) for each reactor is shown in Figure 3.

As shown in Figure 3 increasing HRT has led to increasing the TPH removal efficiency, because pollutants contacted microorganisms for a long hydraulic retention time.

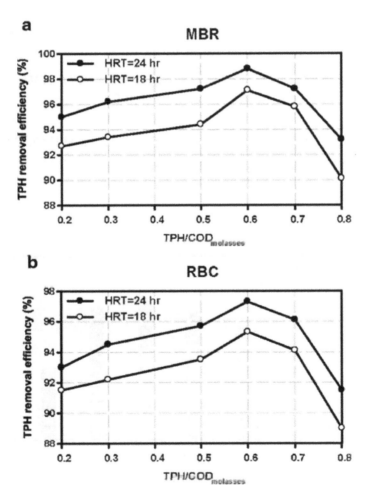

**FIGURE 3:** TPH removal efficiency at two HRTs of 18 and 24 hours versus ratios of TPH/COD molasses in (a) HMBR (b) RBC.

Increasing the ratio of TPH/COD molasses to 0.6 has led to increasing the TPH removal efficiency in both reactors but when the ratios of TPH/COD molasses was greater than 0.6, the efficiency of both systems in removing the pollutant was reduced. This is due to the fact that the increase in the concentration of hydrocarbons on biofilm distorts the cellular metabolism of microorganisms and prevents them from using carbon molasses for their metabolism and reproduction. This will, in turn, reduce MLSS in system and the potential for removing the pollutant will be significantly reduced. Thus, in treating the oily wastewater in such reactors, it is recommended not to choose the ratio of TPH/COD molasses more than 0.6.

### 3.3.3 THE EFFECT OF VARIOUS RATIOS OF TPH/COD MOLASSES ON TPH REMOVAL EFFICIENCY

Figure 4 shows the TPH removal efficiency for ratios of TPH/COD molasses at HRT of 24 hours in both reactors.

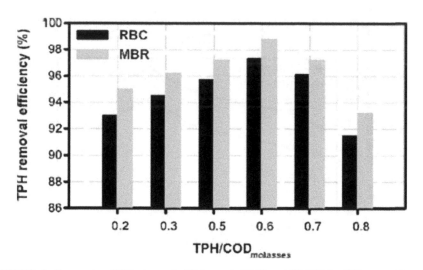

**FIGURE 4:** Comparing TPH removal efficiency in RBC and HMBR in different ratios of TPH/COD molasses at HRT of 24 hours.

This comparison shows that TPH removal efficiency for all concentra-tions of the oily pollutant used in this project has been higher in hybrid membrane than RBC.

### 3.3.4 THE EFFECT OF VARIOUS RATIOS OF TPH/COD MOLASSES ON SUSPENDED SOLIDS REMOVAL EFFICIENCY

Figure 5 shows the suspended solid's removal efficiency by two reactors at various concentrations of the pollutant.

This comparison shows that as the concentration of the oily pollut-ant increases, the suspended solids removal efficiency is reduced in both reactors. The effluent suspended solids of the system was increased with increasing oily pollutant concentrations because the biofilm detached from the media due to the toxicity of oily pollutant [13].

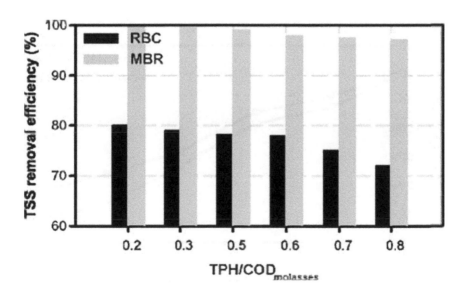

**FIGURE 5:** Comparing the suspended solids' removal efficiency in RBC and hybrid membrane bioreactor for various ratios of TPH/COD at HRT of 24 hours.

Also the diagram shows the higher efficiency of HMBR than the RBC in removing the suspended solids because of the membrane performance.

### 3.3.5 INVESTIGATING THE CHANGES OF PERMEATE FLUX FROM MEMBRANE OVER TIME

Figure 6 shows the changes of permeate flux from membrane in a typical pressure of 1.2 bar.

When the permeate flux of the membrane was about 30 L/m2.hr ( it takes 6 days for average MLSS of 3000mg/l and about 5 days for higher concentrations) chemical cleaning of the membrane is performed.

Higher permeate flux of the membrane at HRT of 24 hours than 18 hours proves higher efficiency of removing organic substances and suspended solids and thus reduction of the membrane fouling and higher permeate flux as well.

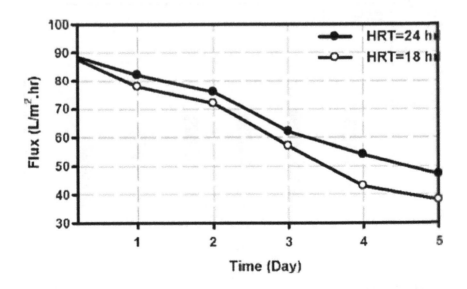

**FIGURE 6:** Permeate flux of membrane versus time.

## 3.4 CONCLUSIONS

In this paper, the behavior of hybrid membrane bioreactor in various load-ings of oily pollutant was studied and the results were compared with the time when the rotating biological contactor performs without using a membrane.

The attached growth bioreactor creates the biofilm on the support me-dia that provide a better treatment efficiency than suspended growth biore-actor due to accumulation of high microbial population in a large surface area. Therefore, better performance can be achieved by combining such a biofilm reactor as RBC with a membrane compared to suspended growth bioreactors as the active sludge in convectional HMBRs.

RBC requires a secondary settling tank which is accomplished by add-ing a membrane to the system. However, it has a smaller volume than the settling tank and the amount of suspended solids in its effluent is less than the effluent from the settling tank. The membrane can also separate the materials that cannot be settled in the settling tank from effluent. It is cost effective as well when there is space limitation or the land is expensive.

Comparison of two hydraulic retention times of 24 and 18 hours for both reactors showed that COD and TPH removal efficiency at 24 hrs HRT is higher than 18 hours.

Results from tests of removing COD and TPH for various ratios of oily pollutant revealed that with the ratio of 0.6 at both HRTs, the amount of COD and TPH removal obtained while with ratios of higher than 0.6, this removal was reduced.

The highest removal efficiency of COD and TPH was 97.3% and 98.8%, respectively. These were obtained by the hybrid membrane biore-actor, with oily pollutant concentration of 700ppm , the ratio of TPH/COD molasses 0.6, at HRT of 24 hours.

The fouling is the major problem with membranes in separation pro-cesses. Nevertheless, RBC was used as a pre-treatment stage and the most of the wastewater was treated before entering into the membrane which results in the reduction of the fouling. Membrane fouling in this study took place after 120 hours from the beginning and after cleaning the membrane was reutilized. This is more than the time needed in previous studies [14].

## REFERENCES

1.  Chih-Ju GJ, Guo-Chiang H: A pilot study for oil refinery wastewater treatment using a fixed-film bioreactor. Adv Environ Res 2003, 7:463-469.
2.  Judd S: The MBR Book: Principles and Applications of Membrane Bioreactors in Water and Wastewater Treatment. Elsevier, Oxford; 2006.
3.  Galil NI, Levinsky Y: Sustainable reclamation and reuse of industrial wastewater including membrane bioreactor technologies: case studies. Desalination 2007, 202:411-417.
4.  Qin J-J, Oo MH, Tao G, Kekre KA: Feasibility study on petrochemical wastewater treatment and reuse using submerged MBR. J Membr Sci 2007, 293:161-166.
5.  Takht RM, Tahereh K, Kargarib A: Application of membrane separation processes in petrochemical industry: a review. Desalination 2009, 235:199-244.
6.  Gander M, Jefferson B, Judd S: Aerobic MBRs for domestic wastewater treatment: a review with cost considerations. Sep Purif Technol 2000, 18:119-130.
7.  Yang W, Cicek N, Ilg J: State-of-the-art of membrane bioreactors: Worldwide research and commercial applications in North America. J Membr Sci 2006, 270:201-211.
8.  Visvanathan C, Aim RB, Parameshwaran K: Membrane separation bioreactors for wastewater treatment. Environ Sci Technol 2000, 30:1-48.
9.  Tawfik A, Temmink H, Zeeman G, Klapwijk B: Sewage treatment in a rotating biological contactor (RBC) system. Water Air Soil Pollut 2006, 175:275-289.
10. Wiszniowski J, Ziembińska A, Ciesielski S: Removal of petroleum pollutants and monitoring of bacterial community structure in a membrane bioreactor. Chemosphere 2011, 83:49-56.
11. Andrew D, Mary Ann H: Standard methods of the examination of water and Wastewater. American Public Health Association, Washington, D.C.; 2005.
12. Pendashteh AR, Fakhru'l-Razi AS, Madaeni S, Abdullah LC, Abidin ZZ, Biak DRA: Membrane foulants characterization in a membrane bioreactor (MBR) treating hyper saline oily wastewater. Chem Eng J 2011, 168:140-150.
13. Akcil A, Karahan AG, Sagdic O: Biological treatment of cyanide by natural isolated bacteria (Pseudomonas sp.). Miner Eng 2003, 16:643-649.
14. Rahman M, Al-Malack M: Performance of a cross flow membrane bioreactor (CF–MBR) when treating refinery wastewater. Desalination 2006, 191:16-26.

# PART II

# METAL INDUSTRIES

Metals enter our rivers, lakes, and oceans from mining, metal finishing, and metal working, threatening the water sources on which the world depends. Although the following articles are limited in scope, copper, nickel, aluminium, lead, zinc, steel, stainless steel, and titanium all play significant roles in our current water crisis. The price humanity pays for the everyday use of minerals is proving to be a high one.

# PART II

# METAL INDUSTRIES

# Removal of Vanadium(III) and Molybdenum(V) from Wastewater Using *Posidonia oceanica* (Tracheophyta) Biomass

CHIARA PENNESI, CECILIA TOTTI, AND FRANCESCA BEOLCHINI

## 4.1 INTRODUCTION

*Posidonia oceanica* (L.) Delile, is the most important and abundant seagrass and it is endemic to the Mediterranean Sea. It forms large underwater meadows from the surface to depths of 40 m, which are an important part of the ecosystem [1, 2] *P. oceanica* has a significant ecological role, as it can form structures known as 'matte', which are monumental constructions that result from the horizontal and vertical growth of the rhizomes with their entangled roots and the entrapped sediment [3]–[5]. This seagrass is very sensitive to human disturbance, such as coastal development, pollution, trawling and high water turbidity [3]–[5]. Indeed, in the year 2000, *P. oceanica* was selected as a Biological Quality Element [6] under the Water Framework Directive [7], as a representative of the aquatic Mediterranean angiosperms for use in the monitoring of the ecological status of coastal waters. *P. oceanica* in Italy is mostly present along the

Removal of Vanadium(III) and Molybdenum(V) from Wastewater Using Posidonia oceanica (Tracheophyta) Biomass. © 2013 Pennesi et al. PLoS ONE 8(10): e76870. doi:10.1371/journal. pone.0076870. Creative Commons Attribution License. Used with the authors' permission.

Tyrrhenian and Ionian coasts, where it is destroyed mainly by trawling and by high water turbidity [8]. Human activities and sea storms result in the accumulation of the leaves of this plant on beaches; and their disposal represents a significant environmental problem [9]. This can, however, be avoided if this waste material can be transformed into a resource.

The physicochemical process known as 'biosorption' indicates the removal of heavy metals from an aqueous solution by passive binding to non-living biomass [10]–[12]. Marine biomass represents an important resource for biosorption processing of heavy metals from industrial wastewater. Indeed, non-living seaweed and seagrass can be used as low-cost sorbents as an alternative to more costly synthetic resins [10]–[16].

The high metal-binding capacity of seaweeds is due to the structure of their cell wall, with various functional groups involved. These include: (1) alginic acid, with carboxyl groups and sulfated polysaccharides, such as fucoidan, and with sulfonic acid, in brown algae matrix (Phylum Ochrophyta) [10], [11], [17]; (2) sulfated galactans such as agar, carregeenan, porphyran, and furcelleran in red algae (Phylum Rhodophyta) [10], [12], [18]; (3) an external capsule that is composed of proteins and/ or polysaccharides in green algae (Phylum Chlorophyta) [19]. While algal materials have been broadly investigated for metal biosorption (Table 1), the potential of marine plants has been notably understudied. However, recent studies in the literature have stated that the ability of seagrasses to adsorb heavy metals also depends on the chemical structure of the plant tissues [11], [12], [20], [21]. In particular, Pennesi at al. [11], [12] studied for the first time the biosorption performance of *Cymodocea nodosa* (Ucria) Ascherson and *Zostera marina* Linnaeus, for the removal of lead and arsenic from aqueous solution. They showed that the optimal biosorption of heavy metal occurs because of the chemical composition of the thin cuticle which is the external layer that covers the leaves. Cutin is a waxy polymer that is the main component of the plant cuticle, and it consists of omega hydroxy acids and their derivatives, which are interlinked via ester bonds, to form a polyester polymer of indeterminate size [22]. This substance is probably responsible for the chemical and physical bond with heavy metals, through the carboxylic groups.

*P. oceanica* not only contains cutin, but it is also a highly fibrous material that is made of cellulose and hemicellulose (ca. 60%–75%) and lignin

**TABLE 1:** The dynamic capacity of LDHs (phenol).

| Macrophyte | Metal | q (mg/g) | C (mg/L) | Conditions | References |
|---|---|---|---|---|---|
| *Fucus vesiculosus* | Cr(III) | 63 | 52–78 | pH 4.5 T25°C | Murphy et al. (2008) |
| | Cr(VI) | 44 | 52–78 | pH 2 | Murphy et al. (2008) |
| | Cd | 73 | - | - | Holan et al. (1993) |
| *Ascophyllum nodosum* | Cu+Pb+Zn+Ni | 117 | 0.5–1 | pH 4 | Zhang and Banks (2006) |
| | Pb | 280 | 200 | pH 6.T25°C | Veglio and Beolchini (1997) |
| | Au | 24 | - | pH 2.5 | Thomas et al. (2003) |
| | Cd | 214 | - | pH 4.9 | Thomas et al. (2003) |
| | Co | 100 | - | pH 4.0 | Thomas et al. (2003) |
| *Fucus spiralis* | Cr(III) | 6165 | 26–52 | pH 4.5 | Murphy et al. (2008) |
| | Cr(VI) | 35.363 | 52–78 | pH 2 | Murphy et al. (2008) |
| | Cd | 146 | - | - | Cordeo et al. (2004) |
| *Ulva lactuca* | Cr(III) | 26.262 | 52–78 | pH 4.5 | Murphy et al. (2008) |
| | Cr(VI) | 2864 | 78–104 | pH 2 | Murphy et al. (2008) |
| *Ulva spp.* | Cr(III) | 5365.2 | 52–78 | pH 4.5 | Murphy et al. (2008) |
| | Cr(VI) | 3064 | 52–78 | pH 2 | Murphy et al. (2008) |
| *Palmaria palmata* | Cr(III) | 3062 | 52–78 | pH 4.5 | Murphy et al. (2008) |
| | Cr(VI) | 3465 | 52–78 | pH 2 | Murphy et al. (2008) |
| *Polysiphonia lanosa* | Cr(III) | 3462 | 52–78 | pH 4.5 | Murphy et al. (2008) |
| | Cr(VI) | 4666 | 52–78 | pH 2 | Murphy et al. (2008) |
| *Laminaria japonica* | Pb(II) | 286 | 300–400 | pH 4.1 | Ghimire et al. (2008) |

**TABLE 1:** CONTINUED..

| | | | | | |
|---|---|---|---|---|---|
| | Cd(II) | 128 | 200–400 | pH 4.1 | Ghimire et al. (2008) |
| | Fe(III) | 49.28 | 100–200 | pH 4.1 | Ghimire et al. (2008) |
| | Ce(III) | 123 | 200–400 | pH 4.1 | Ghimire et al. (2008) |
| *Sargassum fluitans* | Zn | 91 | - | pH 4.5 | Thomas et al. (2003) |
| | Cu | 112 | - | - | Kratochvil et al. (1997) |
| | Cd | 79.52 | - | pH 4.5 | Thomas et al. (2003) |
| *Sargassum natans* | Cd | 135 | - | pH 3.5 T 26°C | Holan et al. (1993) |
| | Au | 400 | 79–120 | pH 2.5 | Thomas et al.(2003) |
| *Sargassum hemiphyllum* | Cr(III) | 72.2 | - | - | Murphy et al. (2008) |
| *Spirogyra* | Cr(III) | 28 | - | - | Murphy et al. (2008) |
| | Pb | 1449 | - | pH 3.5 T 26°C | Murphy et al. (2008) |
| *Halimeda opuntia* | Cr(III) | 40 | - | pH 4.1 T 26°C | Veglio and Beolchini (1997) |
| *Cystoseira compressa* | Pb | 11 | - | room temperature | Pennesi et al. (2012a) |
| *Scytosiphon lomentaria* | Pb | 68 | - | room temperature | Pennesi et al. (2012a) |
| *Ulva rigida* | Pb | 30 | - | room temperature | Pennesi et al. (2012a) |
| *Ulva compressa* | Pb | 46 | - | room temperature | Pennesi et al. (2012a) |
| *Gracilaria bursa-pastoris* | Pb | 19 | - | room temperature | Pennesi et al. (2012a) |
| *Porphyra leucosticta* | Pb | 68 | - | room temperature | Pennesi et al. (2012a) |
| *Polysiphonia sp.* | Pb | 68 | - | room temperature | Pennesi et al. (2012a) |

(ca. 25%–30%), plus a relevant percentage of ash that contains essentially silica and traces of some heavy metals [23]. Moreover, *P. oceanica* contains two type of metallothioneins (MTs) which are a group of cysteine-rich proteins. These proteins have the capacity to bind heavy metals through the thiol groups, which are also known as sulfhydryl groups (R-SH), and amino groups (R-NH2) of its cysteine residues [24], [25]. Other studies have investigated the biosorption of heavy metals such as copper, lead, and chromium, and the removal of dyes from textile waters using *P. oceanica* biomass [21], [26]–[30].

Vanadium and molybdenum are discharged into the environment from various industries that work on alloy steels [31], [32], and these are considered to be persistent environmental contaminants [33], [34]. Due to their toxicity and their accumulation throughout the food chain, they represent a significant problem with ecological and human-health effects. Thus, it is appropriate to eliminate these heavy metals from industrial wastewaters using cheap material such as marine macrophytes [10]–[12], [18], [19], [35].

In the present study, the *P. oceanica* biomass, consisted of dried leaves (material from a beach) that was used for the first time as a low-cost biosorbent for the removal of vanadium(III) and molybdenum(V) from aqueous solution. The objectives of this study were (1) to evaluate the performance of *P. oceanica* non-living biomass in ideal single-metal systems (with either vanadium or molybdenum) under different pH conditions (to determine the optimum operating conditions); (2) to determine any competition phenomena in more real-life systems that were characterized by either high ionic strength or the presence of both vanadium and molybdenum; and (3) to define an equilibrium model to predict the biosorption performance under these more real-life conditions.

## 4.2 MATERIALS AND METHODS

### 4.2.1 SAMPLING AND PREPARATION

The biosorbent materials was obtained from dead leaves of *P. oceanica* (Phylum: Tracheophyta). The samples were collected from Italian beaches of the Ionian Sea: Marina di Leporano, Taranto (Puglia Region; 40°21'43.60"N, 17°20'00.05"E). Specific permission was not required for this geographical location because it is not part of the Marine Protected Area. Furthermore, the material used was destined for waste disposal. After collection, the samples were washed in deionized water (10% w/v, aquaMAXTM, Basic 360 Series) to remove the remaining salt and impurities. Subsequently, the biomass was washed in acid solution (HCl 0.1 N, pH 2; Carlo Erba Reagents) for 4 h under vigorous stirring at room temperature (ratio solid/washing solution 1/10) to remove any traces of metals from the binding sites of the seagrass. Then the leaves were dried (Dry Systems Labconco, Kansas City, MO) at room temperature for 4–5 days to 5 days (to a stable weight) and kept in bottles (Kartell) until use. Before the biosorption tests, each dried biomass sample was reduced into small fragments (of about 0.5 cm, Porcelain Mortar & Pestle Carlo Erba, model 55/8) and re-hydrated before use to increase the percentage removal. This field study did not involve any endangered or protected species.

### 4.2.2 REAGENTS

Stock solutions of molybdenum(V) chloride (MoCl5, Sigma-Aldrich®) and vanadium(III) chloride (VCl3, Sigma-Aldrich®) at, 1 g/L were prepared in distilled water (Sigma-Aldrich®). All of the working solutions at the various concentrations were obtained by successive dilution, according the experimental design (see section 2.5).

### 4.2.3 CHARACTERIZATION OF FUNCTIONAL GROUPS: TITRATION TEST

The characterization of the functional groups on *P. oceanica* that are involved in binding the metals was carried out by acid-base titration test and analyzed according to the Gran Method [36], [37]. Biosorbent materials (i.e, 5 g leaves of *P. oceanica* in 100 mL deionised water) were titrated using standard solutions of NaOH 1 N (basic branch, Sigma-Aldrich®) and H2SO4 0.1 N (acid branch, Sigma-Aldrich®). The pH of the suspension was measured after each addition of titrant (0.05 mL, Eppendorf Research® plus) when stability had been obtained, using a pH-meter (ISteK pH 730p).

## 4.2.4 ADSORPTION TESTS

Before each test, 5 g dried *P. oceanica* biomass was put into 100 mL distilled water for 30 min to rehydrate the sample, with stirring using a magnetic stirrer (MICROSTIRRER-VELP Scientific). The concentrated metal stock solutions (1 g/L molybdenum; 1 g/L vanadium) were added according to the experimental conditions. The suspension pH was adjusted with HCl (0.1 M) and NaOH (0.1 M) and monitored during the whole biosorption test. Aliquot amounts of the solution (1 mL) were periodically sampled to determination the metal concentration(s). The samples were centrifuged (3000 rpm for 5 min.; ALC centrifuge PK 120), to eliminate any suspended matter before the analytical determination, and they were subsequently diluted with acidified water at pH 2 to stabilize the metal(s) before the analytical determinations. Control tests were performed without biomass and showed constant metal concentrations in the solutions over time. This confirmed that precipitation did not take place and that no metal was released by the testing equipment. Metal uptake, q (mg g$^{-1}$), was calculated as the difference in the metal concentration(s) in the aqueous phase before and after sorption, according to Eq. A:

$$q = \frac{V(C_i - C)}{W} \tag{A}$$

where, $V$ is the volume of the solution (L), $C_i$ and $C$ are the initial and equilibrium concentration of metal (molybdenum or vanadium) in solution (mg L$^{-1}$), respectively, and W is the mass of the dry leaves (g). Through regression analysis, the Langmuir adsorption isotherm [38] given in Eq. (B) was adapted to the experimental data:

$$q = \frac{q_{max}C_{eq}}{Ks + C_{eq}} \tag{B}$$

where $q_{max}$ is the maximum adsorption (mg/g) and Ks is the equilibrium constant of the sorption reaction (mg/L).

## 4.2.4 EXPERIMENTAL DESIGN

Table 2 shows the factors and levels investigated for the biosorption of these leaves of *P. oceanica*. All of the experiments were carried out at a constant room temperature. For the ideal systems and the real-life systems with high ionic strength, the equilibrium pH was the only factor considered in the experimental plan. In particular, for the ideal single metal systems, sorption isotherms were evaluated at pH 1, 2, 3, 6, 8, 10 and 12 for vanadium, and at pH 1–3, 5, 7, 9, 10 and 12 for molybdenum. For high ionic strength systems, 20 mg/L NaNO$_3$ was added at the beginning of each experiment, and the sorption isotherms were evaluated at pH 3, 6 and 9 for vanadium, and at pH 6, 9 and 12 for molybdenum,. In the case of the more real-life multi-metal systems, the sorption isotherms were determined at

**TABLE 2:** Factors and levels investigated in the study of molybdenum(V) and vanadium(III) biosorption by biomass of *Posidonia oceanica*.

| Experimental system | Factorial plan | | |
|---|---|---|---|
| | Factors | Levels | |
| Ideal single metal system | Metals | Mo(V) | V(III) |
| | pH | 1–3; 6; 8; 10; 12 | 1–3; 5; 7; 9; 10; 12 |
| Real life high ionic strength (single metal system with 20 mg/L NaNO$_3$) | Metals | Mo(V) | V(III) |
| | pH | 6; 9; 12 | 3; 6; 9 |
| Real life two-metal system | Metal 1 | Mo(V) | V(III) |
| | Metal 2 | absent; 20 mg/L; 40 mg/L | absent; 20 mg/L; 40 mg/L |

pH 3, and the only factor considered in the experimental plan was the presence of the other metal (0,20 and 40 mg/L vanadium/molybdenum for the molybdenum/vanadium sorption, respectively).

## 4.2.5 ANALYTICAL DETERMINATIONS

The pH measurements were made using a pH meter (ISteK 730p). All of the samples were diluted with $HNO_3$ at pH 2 and stored at 4°C before analysis. The metals concentrations in the liquid phase were determined by ICP-AES (Inductively Coupled Plasma Atomic Emission Spectrometry) (Jobin Yvon JY 24, method EPA200.7.2001).

## 4.3 RESULTS AND DISCUSSION

### 4.3.1 BIOMASS CHARACTERIZATION

#### 4.3.1.1 POSIDONIA OCEANICA CHARACTERISATION: ACID-BASE TITRATION

The number and type of functional groups involved in the binding of the metals onto the *P. oceanica* samples were analyzed using acid-base titration [36], [37], [39]. This analysis is based on a neutralization reaction, in order to determine an unknown concentration of functional groups that have an acid behaviour in aqueous solution. The *P. oceanica* titration curve in Figure 1 shows the pH profile as a function of the added meq NaOH: (1) this curve is typical of weak polyprotic acids, which have more than one proton that can be removed by reaction with a base; (2) the equivalence point (i.e., where all of the protons are neutralized by the added hydroxylic groups) is not easily identifiable. The Gran method was used to linearise the titration curve before and after the equivalence point (Fig. S1). The volume necessary to reach the equivalence point was estimated as in the range of 3500–5000 μL (Fig. S1), which when considering the biomass of *P. oceanica* in solution (i.e., 10 g/L), corresponds to a concentration of functional groups with an acid behavior of 3.5–5 m$_{eq}$/g biomass. It can be seen that this titration procedure allowed the estimation of a range for the volume of equivalence, rather than a single value, as in the case of titration of pure acid solutions. This can be explained by considering the heterogeneity of the solid matrix subjected to titration, in terms of these dead leaves of *P. oceanica*. Furthermore, the titration curve in Figure 1 allows an estimation of the acid dissociation constant (K$_a$) of these functional groups that behave as weak acids. Indeed, the pKa (i.e., the pH corresponding to half the equivalence point) [40], is in the range of 3 to 4, which is typical of the R–COO– carboxyl groups of the *P. oceanica* cuticle [17]. Similar results were also obtained for the carboxyl groups in the biomass of the algae *Chlorella pyrenoidosa* H. Chick, *Cyanidium caldarium* (Tilden) Geitler [41] and *Sargassum fluitans* (Børgesen) Børgesen [42]. Moreover, previous studies have used Fourier transform infrared spectroscopy analysis (FTIR) to demonstrate that the functional groups of *P. oceanica* that might have a role in the adsorption process are carboxyl (COOH) and carbonyl stretching (C = O) groups in particular [28], [29].

## 4.3.1.2 HEAVY METAL CONCENTRATION IN POSIDONIA OCEANICA LEAVES

The leaves of *P. oceanica* show significant concentrations of As, Hg, Mo, Pb and V in their tissues (Table 3). Before metals determination, a part of the stock was first washed in acid at pH 2 to remove metals eventually bound at the surface level, while another sample was analyzed under natural conditions. It can be seen that there were no significant differences between the metal concentrations in the two samples (Table 3). This suggests that the *P. oceanica* tissue contains traces of metals, which will be connected to the phenomenon of bioaccumulation, of which there are many studies in the literature [43]–[47]. This result might be related to the presence of industrial sites (e.g., refinery, metallurgical plant and commercial port) located near the collection site (about 30 km away). Indeed, *P. oceanica* is increasingly used as an indicator of chemical contaminations for coastal regions of the Mediterranean Sea, and it is often considered a useful metal bio-indicator [44], [45], [48]. In particular, *P. oceanica* has

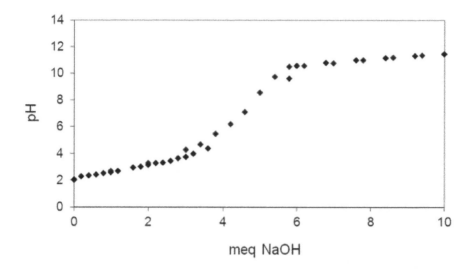

**FIGURE 1:** *P. oceanica* titration curve (biosorbent 10 g/L; room temperature).

been investigated as a bio-indicator for mercury, which was also found in samples used for this study (Table 3), and some studies have suggested that several trace metals can be memorized through analysis of its below-ground tissues [44], [49], [50].

## 4.3.2 VANADIUM AND MOLYBDENUM SPECIATION

In this section, vanadium(III) and molybdenum(V) speciation as a function of pH is discussed, in terms of the predictions of the MEDUSA software [51]. The chemistry of vanadium and molybdenum is complex; indeed, such metals can be present in solution in both anionic and cationic forms. For this reason, theoretical predictions can help both in the choice of experimental conditions to be investigated and in the discussion of the

**TABLE 3:** Heavy metal concentrations in the samples of *Posidonia oceanica* leaves.

| Metal | *P. oceanica* (mg Kg⁻¹) acid wash (pH 2) | *P. oceanica* (mg Kg⁻¹) no acid wash |
|---|---|---|
| arsenic | 2 | 5 |
| mercury | 0.12 | 0.15 |
| molybdenum | 2.5 | 1.6 |
| lead | 2.4 | 3 |
| vanadium | 12 | 14 |

results obtained. Figures 2 and 3 show the predictions for vanadium(III) and molybdenum(V) speciation as a function of pH as given by the ME-DUSA software, where the x-axis indicates the pH while the y-axis shows the logarithm of the metal concentration. It can be seen that anionic forms, cationic forms and insoluble species can be present under different pH conditions (Figs 2, 3). The pH range where vanadium appears to be stable in its ionic form in solution is wide (Fig. 2); indeed, vanadium specia-tion includes both cationic forms (for pH 2–6) and anionic forms (from pH 5 onwards). Consequently, both negatively (e.g., carboxylic groups) and positively (e.g., aminic groups) charged sites of *P. oceanica* biomass are considered to be involved in vanadium biosorption. For molybdenum, Figure 3 shows that it is mainly stable in solution in an anionic form for pH>3; this suggests that positively charged binding sites on the *P. oce-anica* are responsible for molybdenum biosorption.

### 4.3.3 BIOSORPTION: THE IDEAL SINGLE METAL SYSTEM

This part of the study was aimed at an evaluation of the influence of pH on vanadium and molybdenum biosorption by *P. oceanica* in the single metal systems (Table 2; two levels investigated). Figure 4 shows the sorption

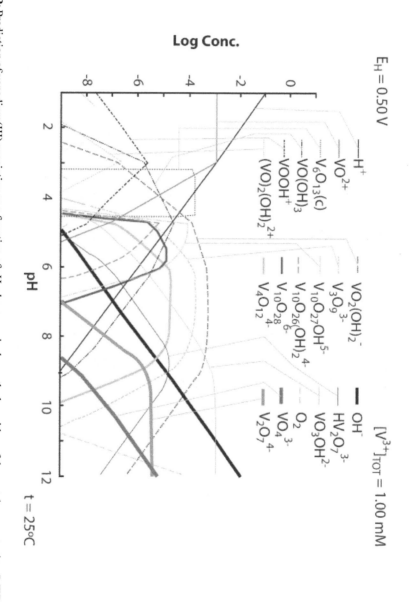

**FIGURE 2:** Prediction of vanadium(III) speciation as a function of pH where y-axis shows the logarithm of the metal concentration (MEDUSA software) [51].

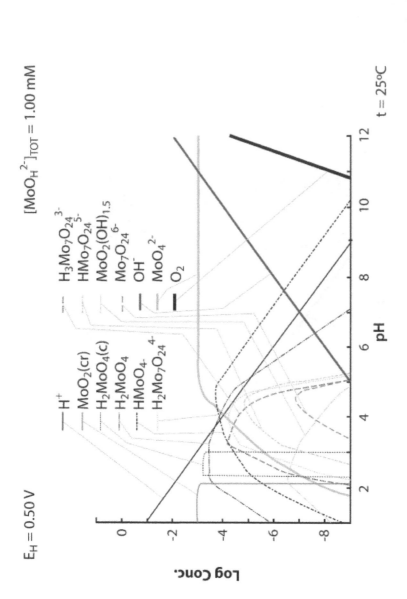

**FIGURE 3:** Prediction of molybdenum(V) speciation as a function of pH where y-axis shows the logarithm of the metal concentration (MEDUSA software) [51].

isotherms for vanadium that were determined for the different equilibrium pHs (pH 1–3, Fig. 4a; pH 3–12, Fig. 4b). It can be seen that the optimal pH for vanadium biosorption by *P. oceanica* is pH 3, with maximum adsorption of about 7 mg/g with an equilibrium concentration of vanadium in solution of 10 mg/L (Fig. 4a). The profiles in Figure 5 also show that the adsorption capacity decreases at both pH<3(Fig. 4a) and >3 (Fig. 4b). Furthermore, control experiments with no biomass excluded any significant precipitation phenomena. For molybdenum biosorption in the single metal ideal systems (Fig. 5), the best adsorption performance was at pH 3 with maximum adsorption around 4 mg/g with the molybdenum equilibrium concentration of 10 mg/L (Fig. 5). Significant precipitation occurred at pH 1 and 2, as also predicted theoretically by the MEDUSA software [51] (Fig. 3). For the pH range from pH 5 to 12, the highest observed values for the specific uptake (q) were around 2 mg/g.

## 4.3.4 BIOSORPTION: REAL-LIFE MULTICOMPONENT SYSTEMS

### 4.3.4.1 HIGH IONIC STRENGTH SYSTEMS

This part of the study was dedicated to the estimation of the influence of pH on vanadium and molybdenum biosorption by *P. oceanica* in the presence of NaNO$_3$ (20 mg/L). Sodium nitrate was added (Table 2) to simulate real-life systems, by providing potential antagonist ions that can compete with the metals for the active sites involved in biosorption on the cuticle of *Posidonia oceanica*. The solution was diluted according to the experimental design (see section 2.5; Table 2; two levels investigated). Figure 6 shows the effects of pH on the sorption isotherms for vanadium and molybdenum, respectively, in the presence of the antagonist ions (NaNO$_3$). A comparison with the data observed in the single metal systems (Fig. 4), confirms that pH 3 was optimal for the adsorption of vanadium (Fig. 6a) and that no competition appears to have taken place. The data in Figure 6b suggest that the adsorption capacity for molybdenum was optimal at pH 12, while in the single metal system, no significant adsorption was observed at this pH (Fig. 5). The oxidative action of NaNO$_3$ might have modified the speciation of molybdenum (Fig. 3), and then modified the

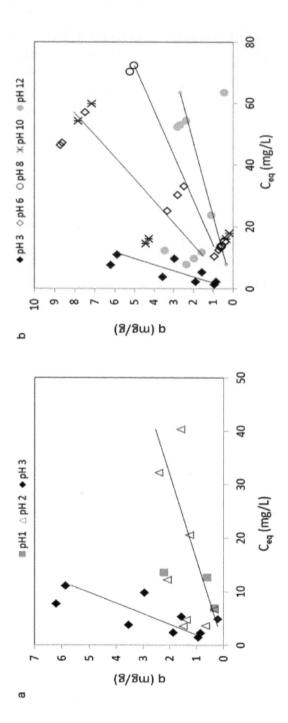

**FIGURE 4:** Sorption isotherms for vanadium in the single metal system, for the range of pH–3 (a) and 3–12 (b) (biosorbent 10 g/L; room temperature).

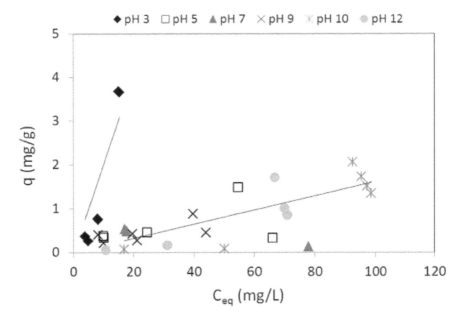

**FIGURE 1:** Sorption isotherms for molybdenum in the single metal system, for the range of pH 3–12 (biosorbent 10 g/L; room temperature).

performances at pH 3. These data show that there are linear correlations between the metal concentrations on the solid and in the liquid in equilibrium, except for the adsorption of vanadium at pH 3, which appears to follow a Langmuir trend.

### 4.3.4.2 MULTI-METAL SYSTEMS.

Potential competition between vanadium and molybdenum might affect the biosorption performances when these metals are present simultaneously. The sorption isotherms were determined at pH 3 considering that this pH was the optimum for the biosorption performances of both of these metals, in the ideal single metal systems. Figure 7 shows the sorption isotherms of each metal in the presence of increasing concentration

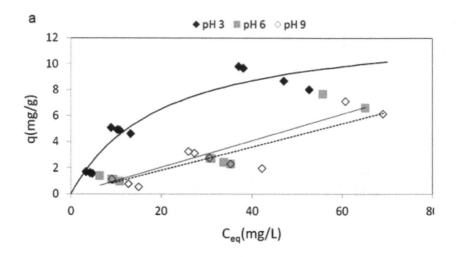

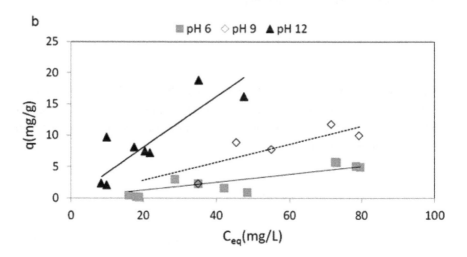

**FIGURE 6:** Sorption isotherms in real life high ionic strength systems: pH effect for vanadium (a) and molybdenum (b) in the presence of $NaNO_3$ 20 mg/L (biosorbent 10 g/L; room temperature).

of the other metal in solutions diluted according to the experimental design (see section 2.5; Table 2; two levels investigated). It can be seen that no competition appeared to take place, and that conversely, the presence of one of the metals appears to favor the adsorption of the other metal. The absence of competition can be explained with reference to theoretical predictions previously reported (see section 3.2): at ca. pH 3, molybdenum is mainly stable as an anion, while vanadium is a cation. Consequently, this indicates that bothof these ions can be adsorbed simultaneously; as they have opposite charges, the two metals interact with different binding sites on the *P. oceanica* cuticle. The Langmuir equation of Eq. (B) was fitted to the experimental data, and Table 4 gives the estimated values for the parameters $q_{max}$ and Ks. Here, it can be seen that parameter $q_{max}$ is estimated at 16 mg/g and 18 mg/g for vanadium and molybdenum, respectively, without any significant effects of the other metal; on the other hand, parameter Ks significantly decreased as the concentration of the other metal increased. This trend confirmed the absence of competition between these two metals, and suggested that the identification of a mathematical model suitable for predicting real multimetal systems should take into account the variability of the sorption equilibrium constant, Ks, depending on the concentration of the other metal. This approach is described in the following section.

**TABLE 4:** Heavy metal concentrations in the samples of *Posidonia oceanica* leaves.

| | vanadium – pH 3 | | | molybdenum – pH 3 | | |
|---|---|---|---|---|---|---|
| | no Mo | 20 mg/L Mo | 40 mg/L Mo | no | V20 mg/L | V40 mg/L V |
| $q_{max}$ (mg/g) | 16 | 16 | 16 | 18 | 18 | 18 |
| $k_s$ (mg/l) | 14 | 8 | 4 | 60 | 34 | 14 |
| $R^2$ | 0.9 | 0.95 | 0.96 | 0.9 | 0.95 | 0.93 |

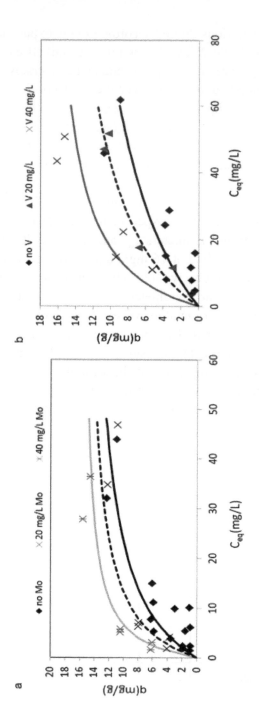

**FIGURE 7:** Sorption isotherms in the two metal system for vanadium (a) and molybdenum (b) in the presence of the other metal respectively (pH 3; biosorbent 10 g/L; room temperature).

### 4.3.4.3 MATHEMATICAL MODELING OF MULTI-METAL EQUILIBRIUM.

As reported above, the theoretical prediction of vanadium speciation (Fig. 2) suggests that under conditions of pH 3, vanadium is substantially present as the $VO_2^+$ cation. Consequently the biosorbent functional groups involved in vanadium sorption are considered to be carboxylic groups (as documented in the literature [28] and shown by the acid-base titration), according to the following simplified sorption/ion-exchange mechanism:

$$R\text{-}COO^- + V_{cation} \rightleftarrows R\text{-}COO^-V_{complex} \qquad (C.1)$$

where the carboxylic groups are dissociated in the water solution according to the following acid dissociation equilibrium:

$$R\text{-}COOH + H_2O \rightleftarrows R\text{-}COO^- + H_3O^+ \qquad (C.2)$$

with a pKA in the range of 3 to 4 as previously reported.

For molybdenum, the theoretical prediction of its speciation (Fig. 3) indicates that at pH 3 it is mainly present as the negatively charged species: H3Mo7O243– and MoO42–, where this latter becomes the ion dominant species as the pH increases. Contrary to what happens for vanadium, in this case, the biosorbent functional groups involved in molybdenum sorption are considered to be positively charged groups, as mainly aminic groups, which are well known as being present on the seagrass cuticle [22], according to the following mechanism:

$$R\text{-}NH_3^+ + Mo_{anion} \rightleftarrows R\text{-}NH_3\text{-}Mo_{complex} \qquad (C.3)$$

where, the aminic groups are dissociated in water solution according to the following acid dissociation equilibrium:

$$R\text{-}NH_3^+ + H_2O \rightleftarrows R\text{-}NH_2 + H_3O^+ \qquad (C.4)$$

with a $pK_A$ in the range 9–10 [52].

In summary, the vanadium and molybdenum sorption process is relatively complex, with many factors involved, such as the metal speciation in solution and the dissociation equilibria of the main functional groups in the biosorbent material. Considering the pKa values reported above, it is expected that under pH 3, at least half of the carboxylic functional groups are available for vanadium biosorption, while all of the aminic groups are positively charged and ready to bind molybdenum complexes. To define a simple mathematical model that can be used to predict the sorption of one metal in the presence of the other, was taken into account that the two metals do not compete for the same sites, due to the opposite charges of the species involved. Consequently, in this case a non-competitive Langmuir equilibrium model can be considered as the simplest one to be tested for data fitting, as follows:

$$q_V = \frac{K_V q_{V,max} C_V}{1 + K_V C_V} \qquad\qquad (C.5)$$

$$q_{Mo} = \frac{K_{Mo} q_{Mo,max} C_{Mo}}{1 + K_{Mo} C_{Mo}} \qquad\qquad (C.6)$$

where $C_V$ and $C_{Mo}$ represent the equilibrium concentrations in solution, $q_V$ and $q_{Mo}$ are the metal specific uptakes, $q_{V,max}$ and $q_{Mo,max}$ are the maximum metal-specific uptakes and $K_V$ and $K_{Mo}$ are the equilibrium constant of sorption in Eqs. (C.1) and (C.3), respectively.

The experimental data reported in Figure 7 shows that for both of these metals, the presence of the other metal favors the biosorption: e.g., vanadium sorption in the presence of molybdenum is higher than in the single metal system. This can be explained by a partial neutralization of charged sites that might have a repulsing action; consequently, in the presence of molybdenum, the positively charged aminic groups are neutralized, and their repulsion towards the positively charged vanadium ions is removed. For molybdenum biosorption an analogous consideration can be performed, as the opposite effect. To take these phenomena into account

**TABLE 5:** Estimated parameters for the synergistic equilibrium sorption isotherms from Eqs. (C.5) and (C.6).

| | |
|---|---|
| $q_{v}$ max | 0.31 mmol/g |
| $K_{V,0}$ | 3.5 L/mmol |
| n | 0.02 |
| $q_{Mo,max}$ | 1.3 mmol/g |
| $K_{Mo,0}$ | 0.14 L/mmol |
| m | 0.12 |
| $R^2$ | 0.89 |

in the sorption model, the presence of one metal increases the equilibrium constant of the other metal sorption according to the following empirical rules:

$$K_V = K_{V,0}(1 + C_{Mo}^n)$$
(C.7)

$$K_{Mo} = K_{Mo,0}(1 + C_V^n)$$
(C.8)

Equations (C.5) and (C.6), with the equilibrium constants given by equations (C.7) and (C.8), were fitted to the experimental data by non linear regression analysis for parameter estimation, through the least-squares method. Table 5 shows the estimated values for the parameters and the performance of the data fitting, which can be considered satisfactory considering that the multiple regression coefficient R2 is near 0.90. Figure 8 shows the equilibrium specific uptake of vanadium (Fig. 8a) and molybdenum (Fig. 8b) as functions of the two metal concentrations, as predicted by equations (C.5) and (C.6), where the synergic effects of the two metals on the biosorption performance can be seen.

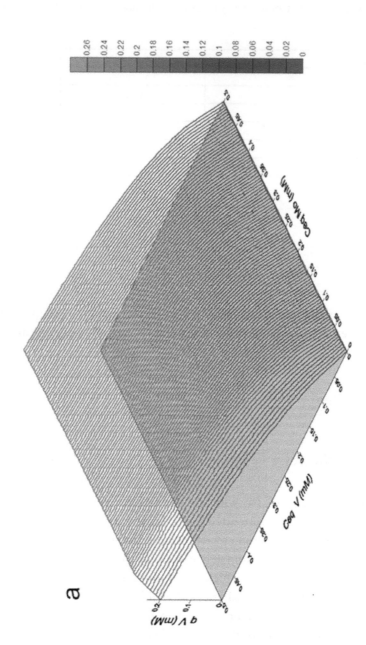

**FIGURE 8:** Sorption isotherms of vanadium (a) and molybdenum (b) as a function of the two metal concentrations, as predicted by equations (C.5) and (C.6) (pH 3, parameters as in Table 5).

FIGURE 8: CONTINUED

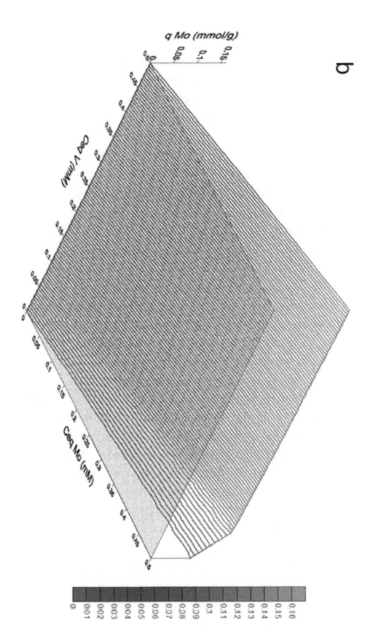

## 4.4 CONCLUSION

The disposal of biomass of *Posidonia oceanica* accumulated on the beaches represents a significant environmental problem [9], that could be avoided if such waste material is transformed as a resource. Some few examples are available in the literature where its potential use as metal biosorbent is assessed: indeed, it has already been demonstrated that *Posidonia oceanica* could adsorb uranium (VI) [53], chromium (VI) [54], lead (II) [28] with the highest sorption capacities of 5.67, 14.48 and 140 mg/g, respectively. Furthermore this biomass has also been reported to effectively adsorb anionic species, such as ortophosphate ions, with sorption capacity of 7.45 mg/g [55]. This work demonstrates the suitability of such biomass also as vanadium and molybdenum sorbent, with a maximum sorption ability estimated at 16 and 18 mg/g, respectively. Sorption of vanadium and molybdenum was explained by chemico-physical interaction (mainly based on ion exchange) with carboxilic and amminic groups that are present in many macromolecules on the cuticle of the plant (e.g. cutin, metallothionein [22], [24], [25]). The real system simulation allowed to exclude any competition phenomena of nitrate ions and of one metal with the other, due to a different speciation of vanadium and molybdenum in aqueous solution (cation vs. anion). This evidence has allowed to develop a new simple multi-metal sorption equilibrium model that is able to take into account the synergic effect on the biosorption performance, that was evident when both metals were present. The availability of a mathematical tool that can predict the performance of biosorption in such multi-metal systems is considered very important [56], [57] and there is a real need that new scientific literature goes beyond the very well-known Langmuir/Freundlich sorption models, representative of ideal single metal systems. Future work will be addressed on one hand at real systems coming from refinery catalysts recycling process [58], [59], on the other hand at the upscale of the biosorption process, in order to find a suitable process configuration for the application of *Posidonia oceanica* biomass at industrial scale.

# REFERENCES

1. Boudouresque CF, Meinesz A (1982) Découverte de l'herbier de Posidonie. Cah Parc nation Port-Cros 4: 1–79.

2. Boudouresque CF, Bernard G, Bonhomme P, Charbonnel E, Diviacco G, et al.. (2006) Préservation et conservation des herbiers a' Posidonia oceanica. Monaco: Ramoge Publication.

3. Piazzi I., Acunto S, Cinelli F (1999) In situ survival and development of Posidonia oceanica (L.) Delile seedlings. Aquatic Botany 63: 103–112. doi: 10.1016/s0304-3770(98)00115-6

4. Francour P, Magréau JF, Mannoni PA, Cottalorda JM, Gratiot J (2006) Management guide for Marine Protected Areas of the Mediterranean sea, Permanent Ecological Moorings. Université de Nice-Sophia Antipolis & Parc National de Port-Cros, Nice.

5. Gacia E, Invers O, Manzanera M, Ballesteros E, Romero J (2007) Impact of the brine from a desalination plant on a shallow seagrass (Posidonia oceanica) meadow. Estuarine, Coastal and Shelf Science 72: 579–590. doi: 10.1016/j.ecss.2006.11.021

6. Med-GIG (2007) WFD Intercalibration technical report for coastal and transitional waters in the Mediterranean ecoregion. WFD Intercalibration Technical Report Part 3: Coastal and Transitional Waters.

7. EC (2000) Directive 2000/60/EC of the European Parliament and the Council of 23 October 2000 establishing a framework for Community action in the field of water policy. Official Journal of the European Community OJ L 327: 1–73. doi: 10.1017/cbo9780511610851.056

8. González-Correa JM, Bayle JT, Sánchez-Lizaso JL, Valle C, Sánchez-Jerez P, et al. (2005) Recovery of deep Posidonia oceanica meadows degraded by trawling. Journal of Experimental Marine Biology and Ecology 320: 65–76. doi: 10.1016/j.jembe.2004.12.032

9. WWF. (2012) Dossier coste "profilo" fragile. Benedetto G, Maceron C, editors.

10. Davis TA, Volesky B, Mucci A (2003) A review of the biochemistry of heavy metal biosorption by brown algae. Water Research 37: 4311–4330. doi: 10.1016/s0043-1354(03)00293-8

11. Pennesi C, Totti C, Romagnoli T, Bianco B, De Michelis I, et al. (2012) Marine Macrophytes as Effective Lead Biosorbents. Water Environment Research 84: 1–8. doi: 10.2175/10.2175/106143011x12989211841296

12. Pennesi C, Vegliò F, Totti C, Romagnoli T, Beolchini F (2012) Nonliving biomass of marine macrophytes as arsenic(V) biosorbents. Journal of Applied Phycology 24: 1495–1502. doi: 10.1007/s10811-012-9808-2

13. Veglio' F, Beolchini F (1997) Removal of metals by biosorption: a review. Hydrometallurgy 44: 301–316. doi: 10.1016/s0304-386x(96)00059-x

14. Volesky B (1990) Removal and Recovery of Heavy Metals by Biosorption. In: Volesky B, editor. Biosorption of Heavy Metals. Florida: CRC Press, Boca Raton. pp.7–43.

15. Davis T, Volesky B, Vieira RHS (2000) Sargassum seaweed as biosorbent for heavy metals. Water Research 34: 4270–4278. doi: 10.1016/s0043-1354(00)00177-9

16. Murphy V, Hughes H, McLoughlin P (2008) Comparative study of chromium biosorption by red, green and brown seaweed biomass. Chemosphere 70: 1128–1134. doi: 10.1016/j.chemosphere.2007.08.015

17. Sheng PX, Ting Y-P, Chen JP, Hong L (2004) Sorption of lead, copper, cadmium, zinc, and nickel by marine algal biomass: characterization of biosorptive capacity and investigation of mechanisms. Journal of Colloid and Interface Science 275: 131–141. doi: 10.1016/j.jcis.2004.01.036

18. Sari A, Tuzen M (2008) Biosorption of total chromium from aqueous solution by red algae (Ceramium virgatum): equilibrium, kinetic and thermodynamic studies. Journal of hazardous materials 160: 349–355. doi: 10.1016/j.jhazmat.2008.03.005

19. Bulgariu D, Bulgariu L (2012) Equilibrium and kinetics studies of heavy metal ions biosorption on green algae waste biomass. Bioresource technology 103: 489–493. doi: 10.1016/j.biortech.2011.10.016

20. Demirak A, Dalman Ö, Tilkan E, Yıldız D, Yavuz E, et al. (2011) Biosorption of 2,4 dichlorophenol (2,4-DCP) onto Posidonia oceanica (L.) seagrass in a batch system: Equilibrium and kinetic modeling. Microchemical Journal 99: 97–102. doi: 10.1016/j.microc.2011.04.002

21. Ncibi MC, Mahjoub B, Hamissa AM Ben, Mansour R Ben, Seffen M (2009) Biosorption of textile metal-complexed dye from aqueous medium using Posidonia oceanica (L.) leaf sheaths: Mathematical modelling. Desalination 109–121. doi: 10.1016/j.desal.2008.04.018

22. Benavente J, Ramos-Barrado J, Heredia A (1998) A study of the electrical behaviour of isolated tomato cuticular membranes and cutin by impedance spectroscopy measurements. Colloids and Surfaces A: Physicochemical and Engineering Aspects 140: 333–338. doi: 10.1016/s0927-7757(97)00290-2

23. Khiari R, Khiari R, Mhenni M, Belgacem M, Mauret E (2010) Chemical composition and pulping of date palm rachis and Posidonia oceanica – A comparison with other wood and non-wood fibre sources. Bioresource Technology 101: 775–780. doi: 10.1016/j.biortech.2009.08.079

24. Giordani T, Natali L, Maserti BE, Taddei S, Cavallini A (2000) Characterization and Expression of DNA Sequences Encoding Putative Type-II Metallothioneins in the Seagrass Posidonia oceanica. Plant Physiology 123: 1571–1582. doi: 10.1104/pp.123.4.1571

25. Cozza R, Pangaro T, Maestrini P, Giordani T, Natali L, et al. (2006) Isolation of putative type 2 metallothionein encoding sequences and spatial expression pattern in the seagrass Posidonia oceanica. Aquatic Botany 85: 317–323. doi: 10.1016/j.aquabot.2006.06.010

26. Ncibi MC, Mahjoub B, Seffen M (2007) Kinetic and equilibrium studies of methylene blue biosorption by Posidonia oceanica (L.) fibres. Journal of Hazardous Materials 139: 280–285. doi: 10.1016/j.jhazmat.2006.06.029

27. Izquierdo M, Gabaldón C, Marzal P, Álvarez-Hornos FJ (2010) Modeling of copper fixed-bed biosorption from wastewater by Posidonia oceanica. Bioresource Technology 101: 510–517. doi: 10.1016/j.biortech.2009.08.018

28. Allouche FN, Mameri N, Guibal E (2011) Pb(II) biosorption on Posidonia oceanica biomass. Chemical Engineering Journal 168: 1174–1184. doi: 10.1016/j.cej.2011.02.005

29. Cengiz S, Tanrikulu F, Aksu S (2012) An alternative source of adsorbent for the removal of dyes from textile waters: Posidonia oceanica (L.). Chemical Engineering Journal 32–40. doi: 10.1016/j.cej.2012.02.015

30. Izquierdo M, Marzal P, Gabald&oacute, n C, Silvetti M, et al. (2012) Study of the Interaction Mechanism in the Biosorption of Copper(II) Ions onto Posidonia oceanica and Peat. CLEAN – Soil, Air, Water 40: 428–437. doi: 10.1002/clen.201100303

31. Staśko R, Adrian H, Adrian A (2006) Effect of nitrogen and vanadium on austenite grain growth kinetics of a low alloy steel. Materials Characterization 340–347. doi: 10.1016/j.matchar.2005.09.012

32. Tang Z, Stumpf W (2008) The role of molybdenum additions and prior deformation on acicular ferrite formation in microalloyed Nb–Ti low-carbon line-pipe steels. Materials Characterization 59: 717–728. doi: 10.1016/j.matchar.2007.06.001

33. Gerke TL, Scheckel KG, Maynard JB (2010) Speciation and distribution of vanadium in drinking water iron pipe corrosion by-products. The Science of the total environment 408: 5845–5853. doi: 10.1016/j.scitotenv.2010.08.036

34. Liber K, Doig LE, White-Sobey SL (2011) Toxicity of uranium, molybdenum, nickel, and arsenic to Hyalella azteca and Chironomus dilutus in water-only and spiked-sediment toxicity tests. Ecotoxicology and environmental safety 74: 1171–1179. doi: 10.1016/j.ecoenv.2011.02.014

35. Montazer-Rahmati MM, Rabbani P, Abdolali A, Keshtkar AR (2011) Kinetics and equilibrium studies on biosorption of cadmium, lead, and nickel ions from aqueous solutions by intact and chemically modified brown algae. Journal of hazardous materials 185: 401–407. doi: 10.1016/j.jhazmat.2010.09.047

36. Michałowski T, Toporek M, Rymanowski M (2005) Overview on the Gran and other linearisation methods applied in titrimetric analyses. Talanta 65: 1241–1253. doi: 10.1016/j.talanta.2004.08.053

37. Michałowski T, Kupiec K, Rymanowski M (2008) Numerical analysis of the Gran methods. A comparative study. Analytica chimica acta 606: 172–183. doi: 10.1016/j.aca.2007.11.020

38. Langmuir I (1918) The Adsorption of Gases on Plane Surfaces of Glass, Mica and Platinum. Journal Of The American Chemical Society 40: 1361–1402. doi: 10.1021/ja02242a004

39. Whittaker G, Mount A, Heal M (2000) BIOS Instant Notes in Physical Chemistry. Taylor, Francis, editors.

40. Shriver D., Atkins PW (1999) Inorganic Chemistry British: Oxford University Press.

41. Gardea-Torresdey JL, Becker-Hapak MK, Hosea JM, Darnall DW (1990) Effect of chemical modification of algal carboxyl groups on metal ion binding. Environmental Science & Technology 24: 1372–1378. doi: 10.1021/es00079a011

42. Fourest E, Serre A, Roux J (1996) Contribution of carboxyl groups to heavy metal binding sites in fungal wall. Toxicological & Environmental Chemistry 54: 1–10. doi: 10.1080/02772249609358291

43. Fourqurean JW, Marbà N, Duarte CM, Diaz-Almela E, Ruiz-Halpern S (2007) Spatial and temporal variation in the elemental and stable isotopic content of the seagrasses Posidonia oceanica and Cymodocea nodosa from the Illes Balears, Spain. Marine Biology 151: 219–232. doi: 10.1007/s00227-006-0473-3

44. Lafabrie C, Pergent G, Pergent-Martini C, Capiomont A (2007) Posidonia oceanica: a tracer of past mercury contamination. Environmental pollution (Barking, Essex: 1987) 148: 688–692. doi: 10.1016/j.envpol.2006.11.015

45. Lafabrie C, Pergent G, Kantin R, Pergent-Martini C, Gonzalez J-L (2007) Trace metals assessment in water, sediment, mussel and seagrass species–validation of the use of Posidonia oceanica as a metal biomonitor. Chemosphere 68: 2033–2039. doi: 10.1016/j.chemosphere.2007.02.039

46. Lafabrie C, Pergent-Martini C, Pergent G (2008) First results on the study of metal contamination along the Corsican coastline using Posidonia oceanica. Marine pollution bulletin 57: 155–159. doi: 10.1016/j.marpolbul.2007.10.007

47. Lafabrie C, Pergent-Martini C, Pergent G (2008) Metal contamination of Posidonia oceanica meadows along the Corsican coastline (Mediterranean). Environmental pollution (Barking, Essex: 1987) 151: 262–268. doi: 10.1016/j.envpol.2007.01.047

48. Ferrat L, Pergent-Martini C, Roméo M (2003) Assessment of the use of biomarkers in aquatic plants for the evaluation of environmental quality: application to seagrasses. Aquatic Toxicology 65: 187–204. doi: 10.1016/s0166-445x(03)00133-4

49. Ferrat L, Gnassia-Barelli M, Pergent-Martini C, Roméo M (2003) Mercury and nonprotein thiol compounds in the seagrass Posidonia oceanica. Comparative biochemistry and physiology Toxicology & pharmacology: CBP 134: 147–155. doi: 10.1016/s1532-0456(02)00220-x

50. Maserti BE, Ferrillo V, Avdis O, Nesti U, Garbo A Di, et al. (2005) Relationship of non-protein thiol pools and accumulated Cd or Hg in the marine macrophyte Posidonia oceanica (L.) Delile. Aquatic Toxicology 75: 288–292. doi: 10.1016/j.aquatox.2005.08.008

51. Puigdomenech I (2009) Program MEDUSA (Make Equilibrium Diagrams Using Sophisticated Algorithms), Department of Inorganic Chemistry.

52. Hoffman RV (2004) Organic chemistry: an intermediate text. 2nd ed. N.Y.: Hoboken, Wiley-Interscience.

53. Aydin M, Cavas L, Merdivan M (2012) An alternative evaluation method for accumulated dead leaves of Posidonia oceanica (L.) Delile on the beaches: removal of uranium from aqueous solutions. Journal of Radioanalytical and Nuclear Chemistry 293, 2: 489–496. doi: 10.1007/s10967-012-1782-2

54. Krika F, Azzouz N, Ncibi MC (2012) Removal of hexavalent chromium from aqueous media using Mediterranean Posidonia oceanica biomass: adsorption studies and salt competition investigation. International Journal of Environmental Research 6, 3: 719–732. doi: 10.1007/s13762-013-0483-x

55. Wahab MA, Hassine RB, Jellali S (2011) Posidonia oceanica (L.) fibers as a potential low-cost adsorbent for the removal and recovery of orthophosphate. Journal of Hazardous Materials 191, 1–3: 333–341. doi: 10.1016/j.jhazmat.2011.04.085

56. Beolchini F, Pagnanelli F, Toro L, Vegliò F (2005) Continuous biosorption of copper and lead in single and binary systems using Sphaerotilus natans cells confined by a membrane: experimental validation of dynamic models. Hydrometallurgy 73–85. doi: 10.1016/j.hydromet.2004.09.003

57. Beolchini F, Pagnanelli F, De Michelis I, Vegliò F (2006) Micellar Enhanced Ultrafiltration for Arsenic(V) Removal: Effect of Main Operating Conditions and

Dynamic Modelling. Environmental Science & Technology 40: 2746–2752. doi: 10.1021/es052114m

58. Beolchini F, Fonti V, Ferella F, Vegliò F (2010) Metal recovery from spent refinery catalysts by means of biotechnological strategies. Journal of Hazardous Materials 178: 529–534. doi: 10.1016/j.jhazmat.2010.01.114

59. Rocchetti L, Fonti V, Vegliò F, Beolchini F (2013) An environmentally friendly process for the recovery of valuable metals form spent refinery catalysts. Waste Management & Research 31, 6: 568–576. doi: 10.1177/0734242x13476364

*There are several supplemental files that are not available in this version of the article. To view this additional information, please use the citation on the first page of this chapter.*

# The Effective Electrolytic Recovery of Dilute Copper from Industrial Wastewater

TENG-CHIEN CHEN, RICKY PRIAMBODO, RUO-LIN HUANG, AND YAO-HUI HUANG

## 5.1 INTRODUCTION

Heavy metals are elements having atomic weights between 63.5 and 200.6 g/mol and a specific gravity greater than 5.0 [1]. With the rapid development of industries such as metal plating facilities, mining operations, fertilizer industries, tanneries, batteries, paper industries, and pesticides, heavy metals wastewaters are directly or indirectly discharged into the environment increasingly, especially in developing countries. Unlike organic contaminants, heavy metals are not biodegradable and tend to accumulate in living organisms, and many heavy metal ions are known to be toxic or carcinogenic. Toxic heavy metals of particular concern in treatment of industrial wastewaters include nickel, mercury, cadmium, lead, and chromium.

Electroplating copper industry discharges huge amount of wastewater into wastewater plant. Copper does essential work in animal metabolism; however, the excessive ingestion of copper brings about serious toxicological concerns, such as vomiting, cramps, convulsions, or even death [2, 3]. Due to the complexity of production processes in electroplating copper manufacturing, only acid washing water, which contains abundant cupric ions, was considered to be reclaimed here. Copper is a really rare and valuable metal. The price of copper is getting higher recently, and the production cost is increasing, and therefore it is really important to consider spent copper plating solution.

In recent years, there is an increasing interest in the development of effective electrochemical methods for the removal of metal ions from wastewaters [4, 5]. The metal ions are effectively recovered from dilute solution using ion exchange technique, but the high cost of resin limits its application [6, 7]; however, the electrolytic process has the advantages of metal recovery without further sequential treatment.

Electrode position has been usually applied for the recovery of metals from wastewater. It is a "clean" technology with no presence of the permanent residues for the separation of heavy metals [8]. Oztekin and Yazicigil (2006) found that electrode position is an applicable method for the recovery of metals under appropriate conditions [9]. They investigated the electrolytic recovery of metals from aqueous solutions containing complexing chelating agents such as EDTA, nitrilotriacetic acid, and citrate in a two-chamber cell separating with a commercial cation-exchange membrane. The results showed that least value of recovery of metal was approximately 40%, and this value increased due to the type of the experiments up to 90% for copper. Chang et al. (2009a) used electro deposition in conjunction with ultrasound to reclaim EDTA-copper wastewater [10]. They found that the technique can effectively remove copper (95.6%) and decompose EDTA (84% COD removal) from wastewater. Issabayeva et al. (2006) [8] presented on the electro deposition of copper and lead ions onto palm shell AC electrodes. Besides, recovery of Cd and Ni by electro deposition was investigated [11].

In this study, we investigate the optimal conditions, including effect of iron concentration, effect of electric current, for electrolytic recovery of Cu+2 ions in simulated copper solution. The optimal experiment condition was applied into the real wastewater form electroplating copper industry, and also the copper recovery efficiency is evaluated in this study.

## 5.2 METHOD AND MATERIALS

### 5.2.1 MATERIALS

The 30000 ppm $CuSO_4$, $FeCl_3 \cdot 6H_2O$, and $FeCl_2 \cdot 4H_2O$ solution was purchased from Merck Company. In order to study the influence of the high organic contaminants on the recovery of copper from the raw effluent, experiments have been carried out in the synthetic effluent without any organic contaminants containing $30000 \, mgL^{-1}$ of Cu in $H_2SO_4$. The reproducibility was checked by performing all the experiments to a minimum of at least three times.

The real industrial wastewaters were supplied form JM chemical and GW industries.

### 5.2.2 METHODS

The experimental setup contains 2 plate electrodes of 18 cm (L) × 17 cm (H) (0.039312 dm2) DSA net as anode and 1 plate electrode of 18 cm (L) × 17 cm (H) ($0.0468 \, dm^2$), stainless steel as cathode. Experiments were performed with 1.4 L copper water solution into a 2.4 L acrylic electrolytic reactor tank (20 cm (L) × 6 cm (W) × 20 cm (H)) with 432 cm² effective cathode surface area. The reactor was also operated at a constant current mode. Recycling pump was also used for the mixing of the solution. The reactor was shown in Figure 1.

The electric power supply was purchased from MaxGood mechanical company (Taiwan) (Figure 2). The effluent volume of 200 mL was taken for all the experiments. Electrolyses were carried out as such in the raw effluent at different current density (0.234, 0.468, 0.585, and 0.702 A/dm2). The copper recovery efficiency effects of no iron, ferrous, and ferric ions were also evaluated in this study.

(a)

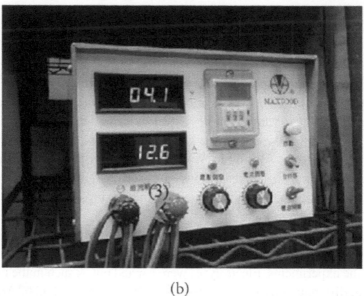

(b)

**FIGURE 1:** The electrical reactor for gold recovery (1) Double electrode cell (2) Recycling pump (3) Power supply.

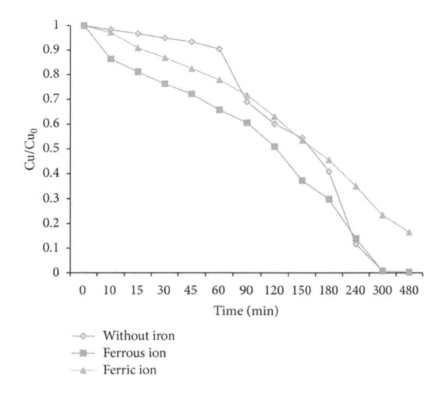

**FIGURE 2:** The copper reduction rate with different types of iron.

In this study, the optimal energy consumption is an important issue. It is important to spend less energy and get higher copper recovery efficiency. The specific energy consumption (EC) is calculated.

## 5.2.3 ANALYSIS METHOD

All the reagents used for chemical precipitation were in industrial quality. All the preparations and experiments were done at the room temperature. All the samples were filtered with TOYO 0.45 μm mixed cellulose ester filters before analysis. Concentrations of Cu and Fe were measured with an atomic absorption spectrophotometer (GBC Sens A), after proper pre-treatment according to the instrument manufacturer's recommendations.

## 5.3. RESULT AND DISCUSSION

### 5.3.1. THE EFFECT OF IRON CONCENTRATION

The behavior of iron during electro-winning of copper from sulphate electrolytes is well known, and its effect on the performance of the copper electro-winning process has been reported by [12–15] who have examined the interaction of iron [Fe+2, Fe+3] during copper electro-winning from dilute solutions, and Cooke et al. (1990) [16] who have studied the mass transfer kinetics of the ferrous-ferric system in copper sulphate electrolyte. Iron effect is an important parameter in the electrical recovery copper system. It is well known that furious is a strong transition metal catalyst, which is applied in the electrical Fenton system [17]. The ferric ion is a strong reduction catalyst; it is reduced in the electrical chemical reaction. The predominant cathode reactions during copper electrowinning from acid sulphate electrolytes can be represented by

Cathode

Copper deposition : $Cu^{+2} + 2e^- \rightarrow Cu$ (1)

Fe (III) reduction : $Fe^{+3} + e^- \rightarrow Fe^{+2}$ (2)

$Cu + 2Fe^{+3} \rightarrow Cu^{+2} + 2Fe^{+2}$ $\Delta E = 0.431$ V (3)

Hydrogen evolution : $2H_3O^- + 2e^- \rightarrow H_2 + 2H_2O$ (4)

Anode

$2H_2O \rightarrow O_2 + 4H^+ + 4e^-$ (5)

$Fe^{+2} \rightarrow Fe^{+3} + e$ (6)

In this research, the three different electrical copper recovery systems are evaluated. The result of copper recovery efficiency between three systems was shown in Figure 3.

Figure 3 and Table 1 showed that the copper reduction efficiency at 480 min reached 83.6%, 99.56%, and 99.9% for ferric, ferrous, and without iron ions, respectively.

Reactions (1) and (2) will be mass transfer controlled during copper electrowinning. Hydrogen evolution will only occur if the cathode current

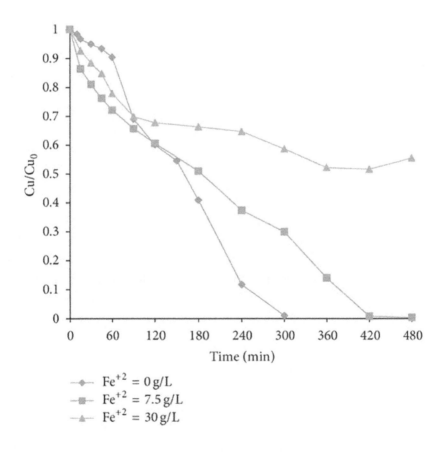

FIGURE 3: The varied for concentration of $Fe^{+2}$ in the electrical copper recovery system.

**TABLE 1:** The copper reduction rate with different types of iron.

| Types of iron | Reduction percent % | | |
|---|---|---|---|
| | 2h | 4h | 8h |
| Without iron | 54.8 | 99 | 99.9 |
| With ferrous ion | 39.5 | 62.8 | 99.56 |
| With ferric ion | 28.3 | 46.6 | 83.6 |

density is greater than the limiting value for copper deposition. Fe+3 was conversed to Fe+2, and also copper metal was also oxidized to Cu+2. The mean effect of raising the Fe+3 concentration caused low copper reduction rate. Moreover, the adverse effect was occurred by increasing the concentration of Fe+3 in the electrical recovery copper system.

According to previous research, adding applicative concentration of Fe+2 will increase the copper recovery efficiency. In this research, different concentrations were evaluated. The result of ferrous recovery was shown in Figure 4. Figure 4 showed that the copper removal efficiency at 300 min was 86.08%, and 54.42% of 7.5 g/L and 30 g/L ferrous ion, respectively. Reactions (5) and (6) show Fe+2 to be oxidized to Fe+3 in the anode. In the cathode, the Fe+3 was reduced to Fe+2, and copper metal was oxidized to copper ion. The result indicated that the concentration of Fe+2 and Fe+3 affected the copper recovery rate in the electrical recovery copper system.

## 5.3.2 EFFECT OF CURRENT DENSITY

Current density is the amount of electrical current flowing in a unit of cross-sectional area of that reactor. Current density is important to the design of electrical and electronic systems. The range of current density was

**FIGURE 4:** The iron effect in the electrical copper recovery system.

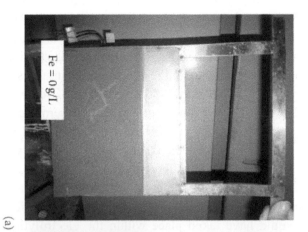

Fe = 0 g/L.

(a)

Fe = 7.5 g/L.

Fe = 30 g/L.

(b)

applied from 0.234 to 0.702 Adm−2 to make a thorough discussion. In the simulation, copper concentration is 30000 mg/L. Figure 5 shows the changes at different current density at pH 2.5 ± 0.3. It is noticed that the rate of copper removal increased with the increased current density. Table 2 and Figure 5 indicated that the copper concentrations removal efficiencies were 20.97%, 86.08%, 98.80%, and 99.90% for 0.234, 0.468, 0.585, and 0.702 Adm−2 at 360 min, respectively. Furthermore, for a longer reaction time of 480 min, the copper concentrations could be down to 28.54 and 19.28 mg L−1 for 0.585 and 0.702 Adm−2, respectively. This observation is better than the results reported in the literature [18].

Depending on the solution composition of the effluent, different reactions can occur at the electrodes. In the present study, following reactions (1)–(6) would have taken place within the electrolytic cell. Due to the optimal economical reason, the current density for 0.585 A/dm2 was the best condition.

### 5.3.3 THE OPTIMAL EXPERIMENTAL CONDITION

From the bench experiment, the $Fe^{+2}$ and $Fe^{+3}$ affected the copper recovery efficiency in the electrical recovery copper system. No iron ion effect was the best condition in this research. In the current density issue, the optimal economical reason, the current density for 0.585 A/dm2 was the best condition.

### 5.4 REAL INDUSTRIAL WASTEWATER

The concnetration of iron and copper ions in the wastewater were analysized by atomic absorption spectrophotometer analysis (AA). The result was shown in Table 3. Table 3 indicated that high concentrations of iron and copper were presented in this real industrial wastewater. Form the manufacture process, the raw materials for the electrical recovery copper system processes are cupric chloride and copper sulfate in the JM chemical and GW industry. Indeed, the optimal experimental conditions were applied into this real wastewater, and this research was evaluated the copper recovery efficiency.

**TABLE 2:** The effect of copper recovery efficiency for current density.

| Times (min) | Current density $(A/dm^2)$ | Copper recover ratio (%) |
|---|---|---|
| 360 | 0.234 | 20.97 |
|  | 0.468 | 86.08 |
|  | 0.585 | 98.80 |
|  | 0.702 | 99.90 |
| 480 | 0.234 | 24.29 |
|  | 0.468 | 99.65 |
|  | 0.585 | 99.9 |
|  | 0.702 | 99.94 |

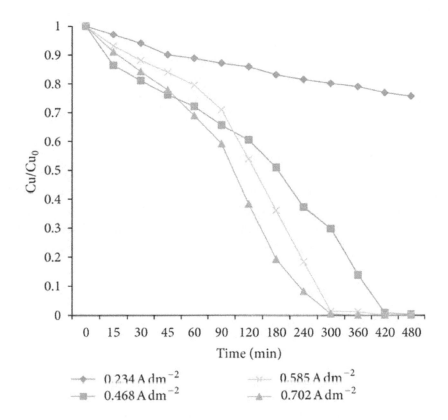

**FIGURE 5:** The copper recovery efficiency for different current density.

TABLE 3: The atomic absorption spectrophotometer (AA) analysis for the Good Weld (GW) and JohgMaw (JM) industrial wastewaters.

|       | [Cu] (ppm) | [Fe] (ppm) |
|-------|------------|------------|
| GW    | 36000      | 50000      |
| JM    | 36000      | 100000     |

## 5.4.1 THE COPPER RECOVERY EFFICIENCY

Figure 6 illustrated the copper recovery efficiency between simulated copper solution and real industrial wastewater. The high copper recovery efficiency occurred, because there is no any iron effect. Table 3 indicated that the wastewater for the iron concentration in the JM chemical was two times higher than Good Weld industry. As reactions (1)–(6), copper metal is oxidized to copper ion, when the high concentration of $Fe^{+3}$ was added. The result was consistent with Das and Krishna research [19].

The chloride ion was an important factor to affect copper recovery efficiency. The electro-derived chlorine chemistry with a DC current could generate several oxidants ($Cl_2$, HOCl, OCl$^-$) in the presence of cupric chloride. The dissolution reactions of copper in chloride solution can be represented as [3, 5, 20].

$$\text{Leach} : Cu + Cl_{2(aq)} \rightarrow Cu^{2+} + 2Cl \qquad (7)$$

$$Cu + Cl_3^- \rightarrow Cu^{2+} + 3Cl^- \qquad (8)$$

$$Cu^{2+} + Cu \rightarrow 2Cu^+ \qquad (9)$$

As a result, one mole of chlorine can oxidize two moles of copper to cuprous state by obtaining two electrons. However, when the dissolution

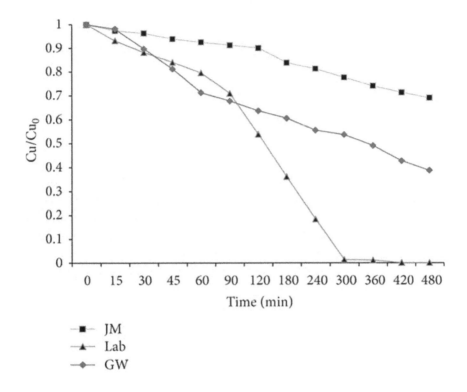

**FIGURE 6:** The copper recovery efficiency between simulated copper solution and the electroplating copper industry wastewater.

rate of chlorine is faster than leaching rate of copper with cupric ions (9), cuprous ions are oxidized with chlorine as follows:

Anode: $Cu^+ \rightarrow Cu^{2+} + e$                                    (10)

Due to the chloride ion effect, the low copper recovery efficiency occurred.

## 5.4.2 THE ENERGY CONSUMPTION (EC)

The specific energy consumption (EC) is calculated from the formula

$$EC = \frac{A(amp) \times V(volt) \times t(hr) \times 1000}{(C_0 - C)(mg/L) \times V_r(L)} \tag{11}$$

where $E \cdot C$ is KWh/Kg; $A$ is operational ampere; $V$ is operational volts; $T$ is operational time; $C$ is initial concentration; $V_r$ is operational volume (L).

Table 4 and Figure 7 show the energy consumption for copper deposition between real industrial wastewater and simulated copper solution. The low energy consumption was around 7.03 KWh/kg for simulated copper solution in the two hours operation time. Due to complex chemical compounds for the real industrial wastewater, such as iron, high energy consumption was around 32.39 KWh/kg. The low copper recovery efficiency form real industrial wastewater was gotten in the electrical recovery copper system.

## 5.5 CONCLUSION

The high copper recovery efficiency can operate for simulated copper solution at high current density (0.585 Adm−2) and obtain copper recovery efficiency (>99.9%). The optimum operational time and current density

**TABLE 4:** Evaluated the energy consumption between real industrial wastewater and simulated copper solution.

| EC | 2 h (KWh/kg) | 4h (KWh/kg) | 6h (KWh/kg) | 8h (KWh/kg) |
|---|---|---|---|---|
| GW | 35.62 | 40.07 | 51.57 | 66.21 |
| JM | 32.39 | 58.34 | 98.59 | 137.93 |
| Lab | 7.03 | 8.03 | 10.02 | 13.27 |

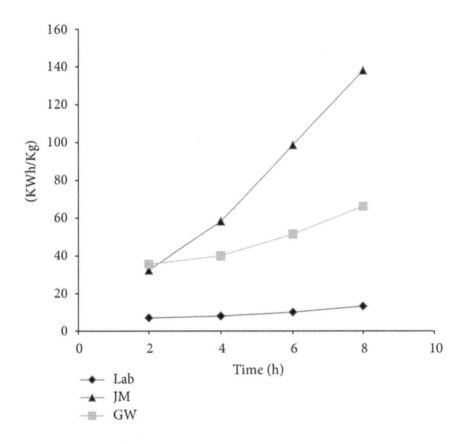

**FIGURE 7:** Evaluated the electrolytic recovery system in the energy consumption for copper recovery between the electroplating copper industry wastewater and simulated copper solution.

were around two hours and 0.585 Adm−2 from simulated copper solution. In the real industrial wastewater, high concentration of iron reduced the copper recovery efficiency; so the high energy consumption (EC) occurred.

Although, the copper recovery efficiency was not high as simulated copper solution, high environmental economical value was included in the technology. In the future, it is also considered that the new type reactor was capable of recovering substantial copper content from industrial wastewater. The possibility of pretreating the wastewater with iron is the necessary step, before the electrical recovery copper system.

## REFERENCES

1.  N. K. Srivastava and C. B. Majumder, "Novel biofiltration methods for the treatment of heavy metals from industrial wastewater," Journal of Hazardous Materials, vol. 151, no. 1, pp. 1–8, 2008.
2.  C. O. Adewunmi, W. Becker, O. Kuehnast, F. Oluwole, and G. Dörfler, "Accumulation of copper, lead and cadmium in freshwater snails in southwestern Nigeria," Science of the Total Environment, vol. 193, no. 1, pp. 69–73, 1996.
3.  A. T. Paulino, F. A. S. Minasse, M. R. Guilherme, A. V. Reis, E. C. Muniz, and J. Nozaki, "Novel adsorbent based on silkworm chrysalides for removal of heavy metals from wastewaters," Journal of Colloid and Interface Science, vol. 301, no. 2, pp. 479–487, 2006.
4.  M. Spitzer and R. Bertazzoli, "Selective electrochemical recovery of gold and silver from cyanide aqueous effluents using titanium and vitreous carbon cathodes," Hydrometallurgy, vol. 74, no. 3-4, pp. 233–242, 2004.
5.  E. Y. Kim, M. S. Kim, J. C. Lee, M. K. Jha, K. Yoo, and J. Jeong, "Effect of cuprous ions on Cu leaching in the recycling of waste PCBs, using electro-generated chlorine in hydrochloric acid solution," Minerals Engineering, vol. 21, no. 1, pp. 121–128, 2008.
6.  T. J. Kanzelmeyer and C. D. Adams, "Removal of copper from a metal-complex dye by oxidative pretreatment and ion exchange," Water Environment Research, vol. 68, no. 2, pp. 222–228, 1996.
7.  K. Fernando, T. Tran, S. Laing, and M. J. Kim, "The use of ion exchange resins for the treatment of cyanidation tailings Part 1 - Process development of selective base metal elution," Minerals Engineering, vol. 15, no. 12, pp. 1163–1171, 2002.
8.  G. Issabayeva, M. K. Aroua, and N. M. Sulaiman, "Electrodeposition of copper and lead on palm shell activated carbon in a flow-through electrolytic cell," Desalination, vol. 194, no. 1–3, pp. 192–201, 2006.
9.  Y. Oztekin and Z. Yazicigil, "Recovery of metals from complexed solutions by electrodeposition," Desalination, vol. 190, no. 1–3, pp. 79–88, 2006.
10. J. H. Chang, A. V. Ellis, C. T. Yan, and C. H. Tung, "The electrochemical phenomena and kinetics of EDTA-copper wastewater reclamation by electrodeposition and ultrasound," Separation and Purification Technology, vol. 68, no. 2, pp. 216–221, 2009.

11. C. C. Yang, "Recovery of heavy metals from spent Ni-Cd batteries by a potentiostatic electrodeposition technique," Journal of Power Sources, vol. 115, no. 2, pp. 352–359, 2003.

12. J. G. G. A. Butts, Copper: The Science and Technology of the Metal, its Alloys and Compounds, vol. 122 of ACS Monograph, Reinhold, New York, NY, USA, 1960.

13. T. N. Anderson, C. N. Wright, and K. J. Richards, "Important electrochemical aspects of electrowinning copper from acid leach solutions," in International Symposium on Hydrometallurgy, D. J. I. Evans and R. S. Shoemaker, Eds., vol. 127, p. 154, AIME, 1973.

14. R. Winand, "Electrocrystallization of copper," Transactions of The Institution of Mining and Metallurgy C, vol. 84, pp. 67–75, 1975.

15. D. W. Dew and C. V. Phillips, "The effect of Fe(II) and Fe(III) on the efficiency of copper electrowinning from dilute acid Cu(II) sulphate solutions with the chemelec cell. Part I. Cathodic and anodic polarisation studies," Hydrometallurgy, vol. 14, no. 3, pp. 331–349, 1985.

16. A. Cooke, J. Chilton, and D. Fray, "Mass-transfer kinetics of the ferrous-ferric electrode process in copper sulphate elecrowinning electrolytes," Transactions of The Institution of Mining and Metallurgy C, vol. 98, 1990.

17. S. Karthikeyan, A. Titus, A. Gnanamani, A. Mandal, and G. Sekaran, "Treatment of textile wastewater by homogeneous and heterogeneous Fenton oxidation processes," Desalination, vol. 281, pp. 438–445, 2011.

18. G. Chen, "Electrochemical technologies in wastewater treatment," Separation and Purification Technology, vol. 38, no. 1, pp. 11–41, 2004.

19. S. C. Das and P. G. Krishna, "Effect of Fe(III) during copper electrowinning at higher current density," International Journal of Mineral Processing, vol. 46, no. 1-2, pp. 91–105, 1996.

20. D. M. Muir, "Basic principles of chloride hydrometallurgy," in Proceedings of the International Conference Chloride Metallurgy Practice and Theory of Chloride/Metal Interaction, E. Peek and G. van Weert, Eds., vol. 2, pp. 759–791, CIM, Montreal, Canada, 2002.

# CHAPTER 6

# Intracellular Biosynthesis and Removal of Copper Nanoparticles by Dead Biomass of Yeast Isolated from the Wastewater of a Mine in the Brazilian Amazonia

MARCIA R. SALVADORI, RÔMULO A. ANDO,
CLÁUDIO A. OLLER DO NASCIMENTO, and BENEDITO CORRÊA

## 6.1 INTRODUCTION

The biosynthesis of NPs is viewed as a new fundamental building pillar of nanotechnology. Nanobiotechnology has revolutionized the production of nanomaterials which are environmentally safe products. Physico-chemical methods employ toxic chemicals and energy intensive routes, which make these choices eco-hazardous and preclude their use for biomedicine and clinical applications [1]. Therefore, environment friendly protocols need to be developed for the synthesis of nanomaterials. Copper NPs have potential industrial applications, including their use as wood preservatives, gas sensors, catalytic processes, high temperature superconductors and

*Intracellular Biosynthesis and Removal of Copper Nanoparticles by Dead Biomass of Yeast Isolated from the Wastewater of a Mine in the Brazilian Amazonia. © 2014 Salvadori et al. PLoS ONE 9(1): e87968. doi:10.1371/journal.pone.0087968. Creative Commons Attribution License (http://creativecommons.org/licenses/by/4.0/) Used with the authors' permission.*

solar cells, among others [2], [3], [4]. The synthesis of different NPs by microorganisms such as prokaryotes (bacteria and actinomicetes) and eukaryotes (yeast, fungi and plant) has been reported in the literature [5]–[6]. Yeasts are preferred for the synthesis of nanomaterials due to their traditional use for bioleaching metals from mineral ores [7]–[8]. Wastewater from copper mining often contains high concentrations of this toxic metal produced during its extraction, beneficiation, and processing. Bioremediation of toxic metals such as copper through biosorption has received a great deal of attention in recent years not only as a scientific novelty, but also because of its potential industrial applications. This approach is competitive, effective, and cheap [9]. In this respect, studies have demonstrated the multi-metal tolerance of *Rhodotorula spp*, which may be of potential use for the treatment heavy metal-bearing wastewater [10]. Consequently, there has been considerable interest in developing methods for the biosynthesis of copper NPs as an alternative to physical and chemical methods. A literature review [11] revealed only few studies on the biosynthesis of copper NPs using fungi and none of the studies has used the yeast *Rhodotorula mucilaginosa* (*R. mucilaginosa*). On the other hand, several studies have investigated the biosynthesis of copper NPs using bacteria, for example, Hasan et al. [12], Ramanathan et al. [13], Singh et al. [14] among others. This work had the objective to enlarge the scope of biological systems for the biosynthesis of copper NPs and bioremediation. We explored for the first time the potential of the yeast *R. mucilaginosa*, for the removal and conversion of copper ions to copper NPs. Thus the goals of uptake and of a natural process to the production of copper NPs, have been achieved in the present study using dead biomass of *R. mucilaginosa*.

## 6.2 MATERIALS AND METHODS

### 6.2.1 ETHICS STATEMENT

The company Vale S.A., owner of Sossego Mine, located in Canaã, Pará, in the Brazilian Amazon region, through the director of the Vale Technology Institute, Dr Luiz Eugenio Mello authorized the establishment and dissemination of the study featured in this research article, allowing the collection of material (water from pond of copper waste) supervised by company employees, whose material led to the isolation of the fungus under study. This field study did not involve manipulation of endangered or protected species by any government agency.

## 6.2.2 GROWTH AND MAINTENANCE OF THE ORGANISM

The yeast *R. mucilaginosa* was isolated from the water collected from a pond of copper waste from Sossego mine, located in Canãa dos Carajás, Pará, Brazilian Amazonia region (06°26′S latitude and 50°4′W longitude). The *R. mucilaginosa* was maintained and activated in YEPD agar medium (10 g yeast extract $L^{-1}$, 20 g peptone $L^{-1}$, 20 g glucose $L^{-1}$ and 20 g agar $L^{-1}$) and the media compounds were obtained from Oxoid (England) [15].

## 6.2.3 ANALYSIS OF COPPER (II) TOLERANCE

Copper tolerance of the isolated yeast was determined as the minimum inhibitory concentration (MIC) by the spot plate method [16]. For this purpose, YEPD agar medium plates containing different copper concentrations (50 to 3000 mg L−1) were prepared and inocula of the tested yeast were spotted onto the metal and control plates (plate without metal). The plates were incubated at 25°C for at least 5 days. The MIC was defined as the lowest concentration of the metal that inhibits visible growth of the isolate.

## 6.2.4 EVALUATION OF COPPER NPS RETENTION BY THE YEAST

### 6.2.4.1 PREPARATION OF THE ADSORBATE SOLUTIONS.

All chemicals used in the present study were of analytical grade and were used without further purification. All dilutions were prepared in double-deionized water (Milli-Q Millipore 18.2 $M\Omega cm^{-1}$ resistivity). The copper stock solution was prepared by dissolving $CuCl_2.2H_2O$ (Carlo Erba, Italy) in double-deionized water. The working solutions were prepared by diluting this stock solution.

## 6.2.4.2 BIOMASS PREPARATION

The yeast cells were grown in 500 mL Erlenmeyer flasks containing 100 mL YEPD broth (10 g yeast extract $L^{-1}$, 20 g peptone $L^{-1}$, 20 g glucose $L^{-1}$). The flasks were incubated on a rotary shaker at 150 rpm for 20 h at 27°C. The biomass was harvested by centrifugation. Once harvested, the biomass was washed twice with double-deionized distilled water and was used directly for the experiment, corresponding to live biomass. For the production of dead biomass, an appropriate amount of live biomass was autoclaved.

## 6.2.4.3 EXPERIMENTAL DESIGN OF THE EFFECTS OF PHYSICO-CHEMICAL FACTORS ON THE EFFICIENCY OF ADSORPTION OF COPPER NPS BY THE YEAST

The effect of pH (2–6), temperature (20–60°C), contact time (5–360 min), initial copper concentration (25–600 mg $L^{-1}$), and agitation rate (50–250 rpm) on the removal of copper was analyzed using analysis of variance models [17] with Bonferroni's multiple comparisons method for adjustment of p-values. These experiments were optimized at the desired pH, temperature, metal concentration, contact time, agitation rate and biosorbent dose (0.05–0.75 g) using 45 mL of a 100 mg $L^{-1}$ of Cu (II) test solution in plastic flask.

Sorption experiments were carried out using several concentrations of copper (II) prepared by appropriate dilution of the copper (II) stock solution. The pH of the solutions was adjusted using HCl or NaOH aqueous solutions. The desired biomass dose was then added and the content of the flask was shaken for the desired contact time in a shaker at the required agitation rate. The reaction mixtures were filtered by vacuum filtration through a Millipore membrane. The filtrate was analyzed for metal concentrations by flame atomic absorption spectrophotometer (AAS). The efficiency (R) of metal removal was calculated using the following equation:

$$R = (C_i - C_e) = C_i. 100$$

where $C_i$ and $C_e$ are initial and equilibrium metal concentrations, respectively. The metal uptake capacity, qe, was calculated using the following equation:

$$q_e = V(C_i - C_e)/M$$

where qe (mg g$^{-1}$) is the biosorption capacity of the biosorbent at any time, M (g) is the biomass dose, and V (L) is the volume of the solution.

### 6.2.4.4 SORPTION ISOTHERMS

The equilibrium data were fitted using the two most commonly adsorption models, Langmuir and Freundlich [18]. The biosorption was analyzed by the batch equilibrium technique using the following sorbent concentrations of 25–600 mg L$^{-1}$. The linearized Langmuir isotherm model is:

$$C_e/q_e = 1/(q_m.b) + C_e/q_m$$

where qm is the monolayer sorption capacity of the sorbent (mg g$^{-1}$), and b is the Langmuir sorption constant (L mg $^{-1}$). The linearized Freundlich isotherm model is:

$$Inq_e = InK_F + 1/n.InC_e$$

where KF is a constant relating the biosorption capacity and 1/n is related to the adsorption intensity of adsorbent.

## 6.2.4.5 BIOSORPTION KINETICS

The experimental biosorption kinetic data were modeled using the pseudo-first-order, and pseudo-second-order models. The linear pseudo-first-order model [19] can be represented by the following equation:

$$\log(q_e - q_t) = \log q_e - K_1 = 2:303_{,}t$$

where, qe (mg g$^{-1}$) and qt (mg g$^{-1}$) are the amounts of adsorbed metal on the sorbent at the equilibrium time and at any time t, respectively, and $K_1$ (min$^{-1}$) is the rate constant of the pseudo-first-order adsorption process. The linear pseudo-second-order model [20] can be represented by the following equation:

$$t/q_t = 1/K_2.q_e^2 + t/q_e$$

where $K_2$ (g mg$^{-1}$ min$^{-1}$) is the equilibrium rate constant of pseudo-second-order.

## 6.2.5 INTRACELLULAR BIOSYNTHESIS OF COPPER NPS BY R. MUCILAGINOSA

Only dead biomass of R. mucilaginosa was used for the analysis of copper NPs production since it exhibited high adsorption capacity of the copper metal ion than live biomass. The biosynthesis of copper NPs by dead biomass of R. mucilaginosa was investigated using the equilibrium data and a solution containing 100 mg L$^{-1}$ copper (II). After reaction with the copper ions, sections of R. mucilaginosa cells were analyzed by transmission electron microscopy (TEM) (JEOL-1010) to determining the size, shape and location of copper NPs on the biosorbent. Analysis of small fragments of the biological material before and after the formation of copper NPs, were performed on pin stubs then coated with gold under vacuum, and examined by SEM (JEOL 6460 LV) equipped with an energy dispersive spectrometer (EDS) to identify the composition of elements of the sample. The XPS analysis was carried out at a pressure of less than $5 \times 10^{-7}$ Pa using a commercial spectrometer (UNI-SPECS UHV System). The Mg Ká line was used (hv = 1253.6 eV) and the analyzer pass energy was set to 10 eV. The inelastic background of the C 1s, O 1s, N 1s and Cu $2p_{3/2}$ electron core-level spectra was subtracted using Shirley's method. The composition (at.%) of the near surface region was determined with an accuracy of $\pm 10\%$ from the ratio of the relative peak areas corrected by Scofield's sensitivity factors of the corresponding elements. The binding energy scale of the spectra was corrected using the C 1s hydrocarbon component of the fixed value of 285.0 eV. The spectra were fitted without placing constraints using multiple Voigt profiles. The width at half maximum (FWHM) varied between 1.2 and 2.1 eV and the accuracy of the peak positions was $\pm 0.1$ eV.

## 6.3 RESULTS AND DISCUSSION

The sensitivity towards at copper of the *R. mucilaginosa* when subjected to minimum inhibitory concentration at different metal concentrations (50–3000 mg $L^{-1}$) showed that this yeast can survive within high level concentrations until 2000 mg $L^{-1}$. The yeast uses several mechanisms to balance intracellular metal concentrations and counter metal toxicity. The resistance mechanism includes sequestration of heavy metals by metallothioneins through their high cysteine content and adsorption of heavy metal cations by the cellular walls [21]–[22].

### 6.3.1 EFFECTS OF THE PHYSICO-CHEMICAL FACTORS ON BIOSORPTION

This study showed that copper removal by *R. mucilaginosa* biomass was significantly influenced by the effects and interactions with the physicochemical factors. As can be seen in Figure 1, the percentage of copper removal was higher for dead biomass than live biomass for all parameters tested (p<0.0001 in all cases). Figure 1A shows that the dead biomass was more efficient in the removal of copper compared with the live biomass (p<0.0001 in the five levels of amount of biosorbent), indicating that dead biomass possess a higher affinity for copper than live biomass. The use of dead biomass for Cu (II) removal has the advantages that it is not toxic and, does not require growth media and nutrients for its maintenance [23]. Therefore the *R. mucilaginosa* may become a potential biosorbent in removing heavy metals from polluted water. The effectiveness of biomass concentration in percentage sorption of the metals was also observed in *Rhodotorula glutinis* [24]. In this study copper biosorption was maximum around pH 5.0, for the two types of biomass (Figure 1B, p<0.0001 in both cases), would be expected to interact more strongly with negatively charged binding sites on the biosorbent. At higher pH levels (pH 5), more ligands with negative charges would be exposed, with the subsequent increase in attraction sites to positively charged metal ion [25]. Some researchers have also investigated the effect of pH on the biosorption of toxic metals and found similar results [26], [27], [28], [29]. The maximum removal of copper was observed at 30°C for the two types of biomass (Figure 1C, p<0.0001 in both cases). The influence of temperature on the sorption of metals has also been reported for the yeast *Pichia stipitis* [28] and *Rhodotorula sp.* Y11 [29], for the bacterium *Morganella pyschrotolerans* [30] and for the plant *Cymbopogon flexuosus* [31] and others. The decrease in adsorption with increasing temperature may be due to the weakening of adsorptive forces between active sites of the adsorbents and the adsorbate species [32]. In Figure 1D, the graph shows sigmoidal kinetics for the types of biomass (p<0.0001 in both cases), which is characteristic of an enzyme-catalyzed reaction. The kinetics of copper NPs formation by dead biomass showed that more than 90% of the particles were formed within 60 min of reaction. The importance of contact time of the metal with the biomass has also been reported for *Rhodotorula sp.* Y11 [33]. The optimum copper removal was observed at an agitation speed of 150 rpm for both types of biomass (Figure 1E, p<0.0001 in both cases). At high agitation speeds, vortex phenomena occur and the suspension is no longer homogenous, a fact impairing metal removal [34]. The percentage of copper adsorption decreased with increasing metal concentration (50–500 mg $L^{-1}$) for both types of biomass, as shown in Figure 1F (p<0.0001 in both cases). The same has been observed for fungi at concentration of Zn ranging from 100–400 mg $L^{-1}$ [35], and for copper removal by *Rhodotorula mucilaginosa* RCL-11 and *Candida sp.* RCL-3 [36].

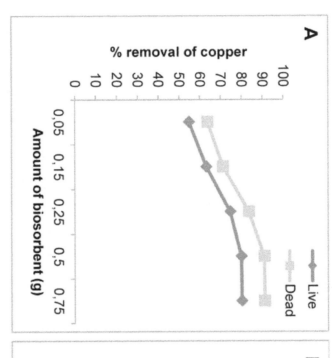

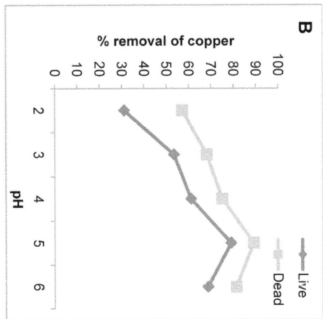

**FIGURE 1:** Sorption studies. Influence of the physico-chemical factors on the live and dead biomass of R. mucilaginosa. (A) Effect of the amount of biosorbent. (B) Effect of pH. (C) Effect of temperature. (D) Effect of contact time. (E) Effect of agitation rate. (F) Effect of initial copper concentration.

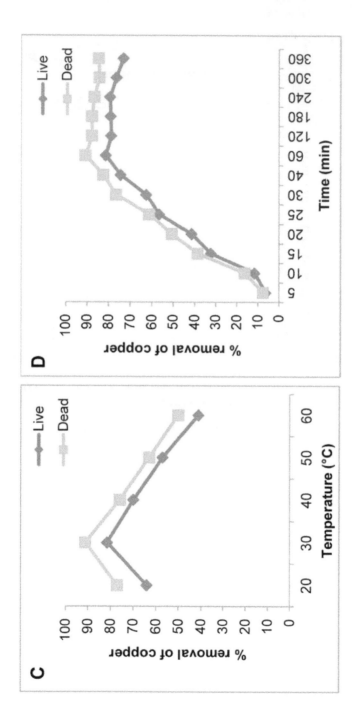

**Figure 1.** CONTINUED.

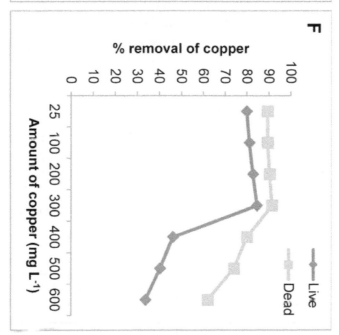
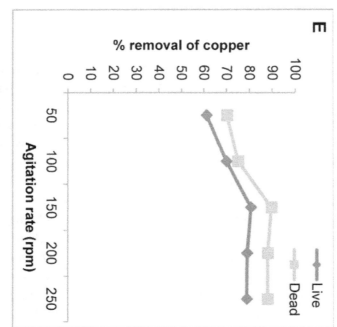

FIGURE 1: CONTINUED.

## 6.3.2 BIOSORPTION ISOTHERMS
## AND ADSORPTION KINETICS MODELS

Langmuir and Freundlich adsorption isotherms, were used to describe the adsorption data for a range of copper (II) concentrations (25–600 mg L$^{-1}$). The Langmuir model better described the Cu (II) biosorption isotherms than the Freundlich model. The Langmuir isotherm for Cu (II) biosorption obtained of the two types of R. mucilaginosa biomass is shown in Figure 2A and Figure 2B. The isotherm constants, maximum loading capacity estimated by the Langmuir and Freundlich models, and regression coefficients are shown in Table 1. The maximum adsorption rate of Cu (II) by R. mucilaginosa (26.2 mg g$^{-1}$) observed in this study was higher than the adsorption rates reported for other known biosorbents, such as Pleurotus pulmonaris, Schizophyllum commune, Penicillium spp, Rhizopus arrhizus, Trichoderma viride, Pichia stiptis, Pycnoporus sanguineus, with adsorption rates of 6.2, 1.52, 15.08, 19.0, 19.6, 15.85 and 2.76 mg g$^{-1}$ respectively [37], [38], [39], [40], [28], [41]. Comparison with biosorbents of bacterial origin showed that the Cu (II) adsorption rate of R. mucilaginosa is comparable to that of Bacillus subtilis IAM 1026 (20.8 mg g$^{-1}$) [42], but higher than the rates reported for the algae Cladophora spp and Fucus vesiculosus (14.28 and 23.4 mg g$^{-1}$) [43]–[44].

TABLE 1: Adsorption isotherm parameters for Cu (II) ions with live and dead biomass of R. mucilaginosa.

| Type of biomass | Langmuir model | | | Freundlich model | | |
|---|---|---|---|---|---|---|
| | $q_m$ (mg g$^{-1}$) | b (L mg$^{-1}$) | R$^2$ | K$_F$ (mg g$^{-1}$) | 1/n | R$^2$ |
| Live | 12.7 | 0.046 | 0.988 | 0.59 | 0.44 | 0.641 |
| Dead | 26.2 | 0.031 | 0.984 | 0.74 | 0.61 | 0.850 |

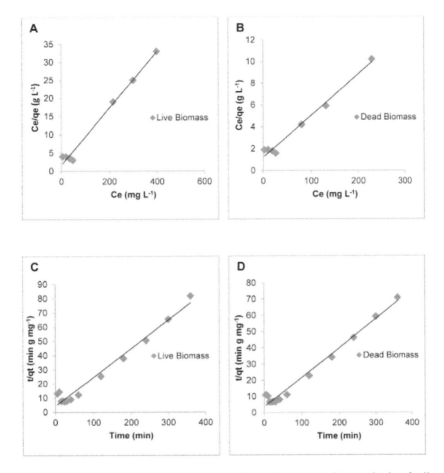

**FIGURE 2:** Equilibrium and kinetic data of the biosorption system. Langmuir plots for live (A) and dead (B) biomass. Pseudo second-order models for live (C) and dead (D) biomass.

Onto both types of biomasses of *R. mucilaginosa* the kinetics of copper biosorption were analysed using pseudo-first-order and pseudo-second-order models. All the constants and regression coefficients are shown in Table 2. In the present study, biosorption by *R. mucilaginosa* was best described using a pseudo-second-order kinetic model as shown

**TABLE 1:** Adsorption isotherm parameters for Cu (II) ions with live and dead biomass of R. mucilaginosa.

| | Langmuir model | | Freundlich model | |
|---|---|---|---|---|
| Type of biomass | $K_1$ (min$^{-1}$) | $R^2$ | $K_2$ (g mg$^{-1}$ min$^{-1}$) | $R^2$ |
| Live | $7.36 \times 10^{-3}$ | 0.474 | $9.45 \times 10^{-3}$ | 0.972 |
| Dead | $6.90 \times 10^{-3}$ | 0.502 | $9.69 \times 10^{-3}$ | 0.981 |

in Figure 2C and Figure 2D. This adsorption kinetics is typical for the adsorption of divalent metals onto biosorbents [45].

## 6.3.3 SYNTHESIS OF COPPER NANOPARTICLES FROM DEAD BIOMASS OF R. MUCILAGINOSA

The researching of biosynthetic methods to the metals NPs formation is important in order to determine even more reliable and reproducible methods for its synthesis and have drawn attention as a simple and viable alternative to chemical procedures and physical methods. The information of the location of copper NPs in the yeast cell is important for elucidating the mechanism of their formation and was obtained through TEM analysis of thin sections of dead biomass (Figure 3). The results clearly showed the high concentration of intracellular copper NPs in the yeast cell, uniformly distributed (monodispersed) without significant agglomeration and was absent in control, the ultrastructural change such as shrinking of cytoplasmatic material was observed in control and in the biomass impregnated with copper due to autoclaving process. However, it was not observed the disruption of the cell wall likely due to the autoclaving method, whose principle consists in causing the death of the microorganisms by denaturation of some proteins [46]. It is important to note, that the cell wall of most yeasts, consists of about 85%–90% polysaccharide and 10%–15% protein and the polysaccharide component consists of a mixture of water-soluble mannan, alkali-soluble glucan, alkali insoluble glucan and small

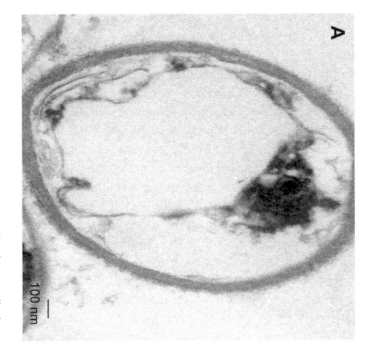

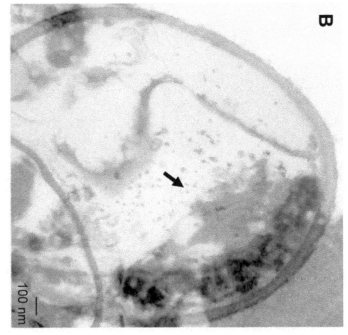

**FIGURE 3:** Transmission electron micrograph of R. mucilaginosa sections. (A) Control (without copper) and (B) Section of the yeast showing intracellular localization of copper NPs (arrow).

amounts of chitin [47], being these components of polysaccharides responsible for the high mechanical resistance of the cell wall [48] (Figure 3A and Figure 3B).

The two most important features that control the chemical, physical, optical and electronic properties of nanoscale materials are the size and shape of these particles [49]–[50]. As observed in Figure 3B, the majority of the particles are spherical in shape and with size of an average diameter of 10.5 nm. To confirm the presence of copper NPs in the dead biomass of yeast it was performed a spot profile SEM-EDS measurement. SEM micrographs recorded before and after biosorption of Cu (II) by yeast biomass was showed in Figure 4A and Figure 4B respectively. It was observed a surface modification by an increasing of the irregularity, after binding of copper NPs with the yeast biomass. The EDS spectrum recorded in the examined region of the yeast cells confirmed the presence of copper. (Figure 5A and Figure 5B). The signals for C, N, O and P may be originate from biomolecules that are bound to the surface of copper NPs.

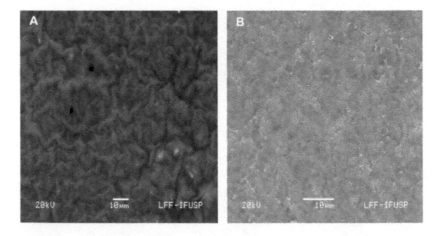

**FIGURE 4:** SEM-EDS analysis of the surface of dead biomass of R. mucilaginosa.(A) Before adsorption of copper ion and (B) after adsorption of copper ion.

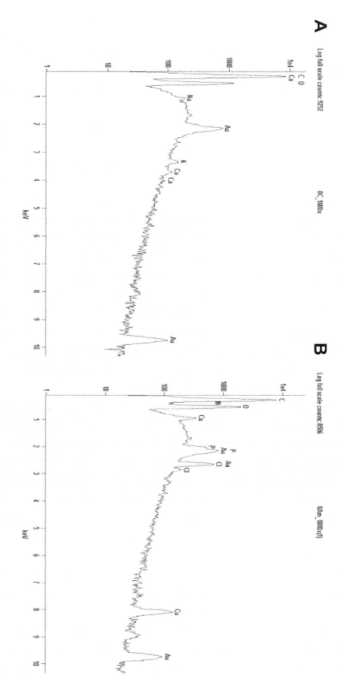

**FIGURE 5:** EDS spectra of dead biomass of R. mucilaginosa. (A) before exposure to copper solution and (B) after exposure to the metal confirming the presence of copper.

Unfortunatly the intracellular mechanism formation of copper NPs by the dead biomass of yeast *R. mucilaginosa* is not fully understood at the moment. The copper ions possibly could diffuse through the cell wall and are reduced by enzymes present on the cytoplasmic membrane and within the cytoplasm. This enzyme-based pathway, also was proposed to silver nanoparticles synthesis [51]. The XPS spectra Figure 6 shows C 1s, N 1s, O 1s and Cu 2p core level after biosynthesis of copper NPs by dead biomass of *R. mucilaginosa*. As can be seen in the high resolution spectra of carbon (C 1s), the components of higher binding energy were deconvoluted within four elements. The main component at 284.8 eV is attributed to the hydrocarbon chains of the celular phase; the peak at 286.7 eV to the α- carbon, the peak at 288.0 eV to the carbonyl groups, and finally the peak at 289.2 eV to the carboxylic groups from the peptides/proteins bound to copper NPs [52]. The deconvoluted spectra of oxygen (O 1s), showed peaks at 531.2 eV, 532.4 eV and 533.3 eV related to the peaks found in the C 1s spectra. The spectra of nitrogen (N 1s) have two components, the main 400.1 eV and a lower at 402.3 eV. In the O 1s and N 1s spectra the major binding energies at 532.4 eV and 400.1 eV respectively were observed confirming the presence of proteins involving copper NPs, which suggests the possibility of these agents acting as capping agents [52]. The Cu 2p core level showed a sharp peak arise at 932,9 eV and it corresponds to the Cu 2p3/2 level characteristic of Cu(0) [53], [54], [55]. The presence of CuO (Cu (II)) phase can be excluded considering the lacking of the signal at 933.7 eV, as well as the presence of Cu2O, that can be is ruled out by the fact of no Cu 2p satellite peak appears with Cu2O [56]–[57].There are several reports in literature of yeast mediating the synthesis of nanoparticles of metal ions except to copper, such as peptide-bound CdS quantum crystallites by *Candida glubrata* [58], *Schizosaccharomyces pombe* also produced CdS nanoparticles [59], PbS nanocrystallites by *Torulopsis sp.* [60], gold nanoparticles by *Pichia jadinii (Candida utilis)* [61]–[62], the tropical marine yeast *Yarrowia lipolytica* NCIM 3589 also synthesized gold nanoparticles [63], Sb2O3 nanoparticles by *Saccharomyces cerevisiae* [64] and silver nanoparticles by yeast MKY3 [65]. Honary et al. [66] reported the production of copper nanoparticles by filamentous fungi, but the authors only used live biomass. The bioprocess proposed here,

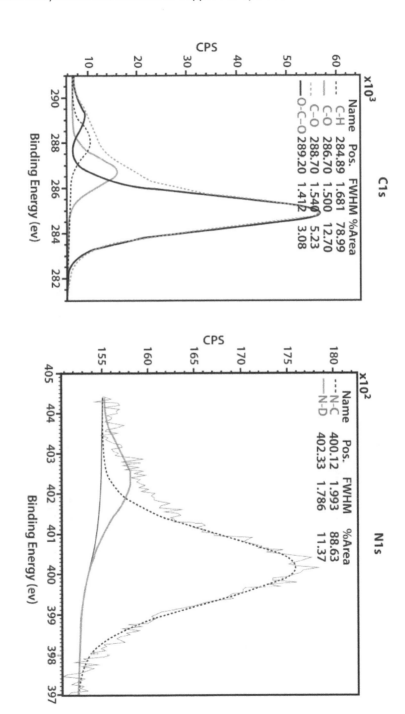

**FIGURE 6:** XPS spectra. C 1s, N 1s, O 1s and Cu 2p core level binding energies after biosynthesis of copper NPs.

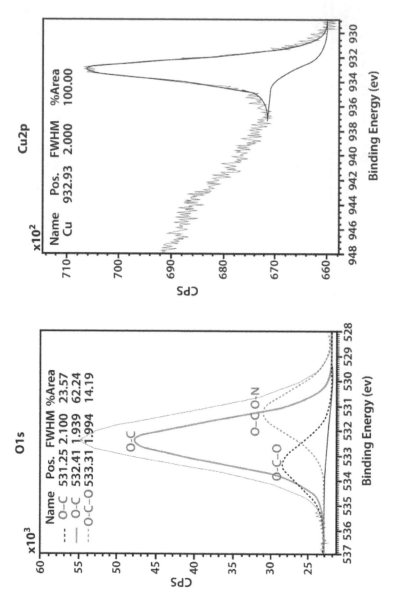

**FIGURE 6:** CONTINUED

using dead biomass has the advantages, that it is not toxic, and does not require growth media and nutrients for its maintenance. The intracellular production of copper NPs by dead biomass of *R. mucilaginosa* was probably the result of a reduction process inside the cell mediated by proteins and enzymes present in the cytoplasm. However the type of proteins interacting with copper NPs remains to be determined and this knowledge would open new perspectives for a more efficient green synthesis of copper NPs.

## 6.4 CONCLUSIONS

In summary, we described for the first time a biological economic template, and non-toxic using dead biomass of the red yeast *R. mucilaginosa*, that may be considered an efficient bioprocess to the synthesis of metallic copper NPs, and probable system for the adsorption of copper ions from wastewater. The dead biomass had a dual role, acting as a reducing agents and stabilizer during the formation of copper NPs, as well as in the uptake of copper ions during the bioremediation process. This natural method affords an amenable process to large scale commercial production. In future studies, we intend to characterize the role of biomacromolecules in the biosorption and bioreduction processes during the synthesis of copper NPs.

## REFERENCES

1. Singh AV, Patil R, Anand A, Milani P, Gade WN (2010) Biological synthesis of copper oxide nanopaticles using Escherichia coli. Curr Nanosci 6: 365–369. doi: 10.2174/157341310791659062
2. Evans P, Matsunaga H, Kiguchi M (2008) Large-scale application of nanotechnology for wood protection. Nat Nanotechnol 3: 577. doi: 10.1038/nnano.2008.286
3. Li Y, Liang J, Tao Z, Chen J (2007) CuO particles and plates: Synthesis and gas-sensor application. Mater Res Bull 43: 2380–2385. doi: 10.1016/j.materresbull.2007.07.045
4. Guo Z, Liang X, Pereira T, Scaffaro R, Hahn HT (2007) CuO nanoparticle filled vinyl-ester resin nanocomposites: Fabrication, characterization and property analysis. Compos Sci Tech 67: 2036–2044. doi: 10.1016/j.compscitech.2006.11.017

5.  Bharde AA, Parikh RY, Baidakova M, Jouen S, Hannoyer B, et al. (2008) Bacteria-mediated precursor-dependent biosynthesis of superparamagnetic iron oxide and iron sulfide nanoparticles. Langmuir 24: 5787–5794. doi: 10.1021/la704019p

6.  Lang C, Schüler D, Faivre D (2007) Synthesis of magnetite nanoparticles for bio- and nanotechnology: genetic engineering and biomimetics of bacterial magneto-somes. Macromol Biosci 7: 144–151. doi: 10.1002/mabi.200600235

7.  Kroger N, Deutzmann R, Sunper M (1999) Polycationic peptides from diatom bio-silica that direct silica nanosphere formation. Science 286: 1129–1132. doi: 10.1126/science.286.5442.1129

8.  He SY, Zhang Y, Guo Z, Gu N (2008) Biological synthesis of gold nanowires us-ing extract of Rhodopseudomonas capsulata. Biotechnol Prog 24: 476–480. doi: 10.1021/bp0703174

9.  Volesky B (2001) Detoxification of metal bearing effluents: biosorption for the next century. Hydrometallurgy 59: 203–216. doi: 10.1016/s0304-386x(00)00160-2

10. Li Z, Yuan H, Hu X (2007) Cadmium-resistance in growing Rhodotorula sp. Y11. Bioresource Technol 99: 1339–1344. doi: 10.1016/j.biortech.2007.02.004

11. Varshney R, Bhadauria S, Gaur MS (2012) A review: Biological synthesis of sil-ver and copper nanoparticles. Nano Biomed Eng 4: 99–106. doi: 10.5101/nbe.v4i2.p99-106

12. Hasan SS, Sing S, Parikh RY, Dharne MS, Patole MS, et al. (2008) Bacterial syn-thesis of copper/copper oxide nanoparticles. J Nanosci Nanotechnol 8: 3191–3196. doi: 10.1166/jnn.2008.095

13. Ramanathan R, Field MR, O'Mullane AP, Smooker PM, Bhargava SK, et al. (2013) Aqueous phase synthesis of copper nanoparticles: a link between heavy metals re-sistance and nanoparticle synthesis ability in bacterial systems. Nanoscale 5: 2300–2306. doi: 10.1039/c2nr32887a

14. Singh V, Patil R, Ananda A, Milani P, Gade W (2010) Biological synthesis of cop-per oxide nanoparticles using Escherichia coli. Curr Nanosci 6: 365–369. doi: 10.2174/157341310791659062

15. Machado MD, Soares EV, Soares HMVM (2010) Removal of heavy metals using a brewer's yeast strain of Saccharomyces cerevisiae: Chemical Speciation as a tool in the prediction and improving of treatment efficiency of real electroplating effluents. J Hazard Mater 180: 347–353. doi: 10.1016/j.jhazmat.2010.04.037

16. Ahmad I, Ansari MI, Aqil F (2006) Biosorption of Ni, Cr and Cd by metal tolerante Aspergillus niger and Penicillium sp using single and multi-metal solution. Indian J Exp Biol 44: 73–76.

17. Neter J, Kutner MK, Nachtsheim CJ, Wasserman W (1996) Applied linear statistical models. 4rd edn, Irwin: Chicago.

18. Volesky B (2003) Biosorption process simulation tools. Hydrometallurgy 71: 179–190. doi: 10.1016/s0304-386x(03)00155-5

19. Lagergren S (1898) About the theory of so called adsorption of soluble substances. Kung Sven Veten Hand 24: 1–39.

20. Ho YS, Mckay G (1999) Pseudo-second-order model for sorption process. Process Biochem 34: 451–465. doi: 10.1016/s0032-9592(98)00112-5

21. Gomes DS, Fragoso LC, Riger CJ (2002) Regulation of cadmium uptake by Saccharomyces cerevisiae. Biochem Biophys Acta 1573: 21–25. doi: 10.1016/s0304-4165(02)00324-0

22. Cho DH, Kim EY (2003) Characterization of Pb2+ biosorption from aqueous solution by Rhodotorula glutinis. Bioproc Biosyst Eng 25: 271–277.

23. Bishnoi NR (2005) Garima (2005) Fungus – An alternative for bioremediation of heavy metal containing wastewater: A review. J Sci Ind Res 64: 93–100.

24. Cho DH, Kim EY (2002) The mechanisms of Pb2+ removal from aqueous solution by Rhodotorula glutinis. Theories and Applications of Chem Eng 8: 4037–4040.

25. Selatnia A, Boukazoula A, Kechid N, Bakti MZ, Chergui A, et al. (2004) Biosorption of lead (II) from aqueous solution by a bacterial dead Streptomyces rimosus biomass. Biochem Eng J 19: 127–135. doi: 10.1016/j.bej.2003.12.007

26. Fourest E, Roux JC (1992) Heavy metal biosorption by fungal micelial by-products: mechanisms and influence of pH. Appl Microbiol Biotechnol 37: 399–403. doi: 10.1007/bf00211001

27. Göksungur Y, Üren S, Güvenc U (2005) Biosorption of cadmium and lead ions by ethanol treated waste baker's yeast biomass. Bioresource Technol 96: 103–109. doi: 10.1016/j.biortech.2003.04.002

28. Yilmazer P, Saracoglu N (2009) Bioaccumulation and biosorption of copper (II) and chromium (III) from aqueous solutions by Pichia stiptis yeast. J Chem Technol Biot 84: 604–610. doi: 10.1002/jctb.2088

29. Li Z, Yuan H (2006) Characterization of cadmium removal by Rhodotorula sp. Y11. Appl Microbiol Biotechnol 73: 458–463. doi: 10.1007/s00253-006-0473-8

30. Ramanthan R, O'Mullane AP, Parikh RY, Smooker PM, Bhargava SK, et al. (2011) Bacterial kinetics-controlled shape-directed biosynthesis of silver nanoplates using Morganella pyschrotolerans. Langmuir 27: 714–719. doi: 10.1021/la1036162

31. Rai A, Singh A, Ahmad A, Sastry M (2006) Role of Halide ions and temperature on the morphology of biologically synthesized gold nanotriangles. Langmuir 22: 736–741. doi: 10.1021/la052055q

32. Pandey KK, Prasad G, Singh VN (1986) Use of wallastonite for the treatment of Cu (II) reach effluents. Water Air Soil Pollut 27: 287–296. doi: 10.1007/bf00649410

33. Li Z, Yuan H, Hu X (2008) Cadmium-resistance in growing Rhodotorula sp. Y11. Bioresource Technol 99: 1339–1344. doi: 10.1016/j.biortech.2007.02.004

34. Liu YG, Fan T, Zeng GM, Li X, Tong Q, et al. (2006) Removal of cadmium and zinc ions from aqueous solution by living Aspergillus niger. Trans Nonferrous Met Soc China 16: 681–686. doi: 10.1016/s1003-6326(06)60121-0

35. Faryal R, Lodhi A, Hameed A (2006) Isolation, characterization and biosorption of zinc by indigenous fungal strains Aspergillus fumigatus RH05 and Aspergillus flavus RH07. Pak J Bot 38: 817–832.

36. Villegas LB, Amoroso MJ, De Figueroa LIC (2005) Copper tolerant yeasts isolated from polluted area of Argentina. J Basic Microbiol 45: 381–391. doi: 10.1002/jobm.200510569

37. Veit MT, Tavares CRG, Gomes-da-Costa SM, Guedes TA (2005) Adsorption Isotherms of copper (II) for two species of dead fungi biomasses. Process Biochem 40: 3303–3308. doi: 10.1016/j.procbio.2005.03.029

38. Du A, Cao L, Zhang R, Pan R (2009) Effects of a copper-resistant fungus on copper adsorption and chemical forms in soils. Water Air Soil Poll 201: 99–107. doi: 10.1007/s11270-008-9930-6

39. Rome L, Gadd DM (1987) Copper adsorption by Rhizopus arrhizus, Cladosporium resinae and Penicillium italicum. Appl Microbiol Biotechnol 26: 84–90. doi: 10.1007/bf00282153

40. Kumar BN, Seshadri N, Ramana DKV, Seshaiah K, Reddy AVR (2011) Equilibrium, Thermodynamic and Kinetic studies on Trichoderma viride biomass as biosorbent for the removal of Cu (II) from water. Separ Sci Technol 46: 997–1004. doi: 10.1080/01496395.2010.537727

41. Yahaya YA, Matdom M, Bhatia S (2008) Biosorption of copper (II) onto immobilized cells of Pycnoporus sanguineus from aqueous solution: Equilibrium and Kinetic studies. J Hazard Mater 161: 189–195. doi: 10.1016/j.jhazmat.2008.03.104

42. Nakajima A, Yasuda M, Yokoyama H, Ohya-Nishiguchi H, Kamada H (2001) Copper sorption by chemically treated Micrococcus luteus cells. World J Microb Biot 17: 343–347.

43. Elmacy A, Yonar T, Özengin N (2007) Biosorption characteristics of copper (II), chromium (III), nickel (II) and lead (II) from aqueous solutions by Chara sp and Cladophora sp. Water Environ Res 79: 1000–1005. doi: 10.2175/106143007x183961

44. Grimm A, Zanzi R, Björnbom E, Cukierman AL (2008) Comparison of different types of biomasses of copper biosorption. Bioresource Technol 99: 2559–2565. doi: 10.1016/j.biortech.2007.04.036

45. Reddad Z, Gerent C, Andres Y, LeCloirec P (2002) Adsorption of several metal ions onto a low-cost biosorbents: kinetic and equilibrium studies. Environ Sci Technol 36: 2067–2073. doi: 10.1021/es0102989

46. Tortora GJ, Funke BR, Case CL (1998) Microbiology an Introduction. 6rd edn. California: Addison Wesley Longman.

47. Nguyen TH, Fleet GH, Rogers PL (1998) Composition of the cell walls of several yeast species. Appl Microbiol Biotechnol 50: 206–212. doi: 10.1007/s002530051278

48. Nimrichter L, Rodrigues ML, Rodrigues EG, Travassos LR (2005) The multitude of targets for the immune system and drug therapy in the fungal cell wall. Microbes Infect 7: 789–798. doi: 10.1016/j.micinf.2005.03.002

49. Alivisatos AP (1996) Perspectives on the physical chemistry of semiconductor nanocrystals. J Phys Chem 100: 13226–13239. doi: 10.1021/jp9535506

50. Aizpurua J, Hanarp P, Sutherland DS, Käll M, Bryant GW, et al. (2003) Optical properties of gold nanorings. Phys Rev Lett 90: 57401–57404. doi: 10.1103/physrevlett.90.057401

51. Sanghi R, Verma P (2009) Biomimetic synthesis and characterization of protein capped silver nanoparticles. Bioresource Technol 100: 501–504. doi: 10.1016/j.biortech.2008.05.048

52. Bansal V, Ahamad A, Sastry M (2006) Fungus-mediated biotransformation of amorphous silica in rice husk to nanocrystalline Silica. J Am Chem Soc 128: 14059–14066. doi: 10.1021/ja062113+

53. Naumkin AV, Kraut-Vass A, Gaarenstroom SW, Powell CJ (2012) NIST X-ray Photoelectron Spectroscopy Database: NIST Standard Reference Database 20, v. 4.1. Available: http://www.srdata.nist.gov/XPS/. Accessed 03 December 2013.

54. Briggs D, Seah MP (1990) Pratical Surface Analysis, Auger and X-ray Photoelectron Spectroscopy. Vol. 1. United Kingdom: John Wiley & Sons, Chichester.

55. Jeong S, Woo K, Kim D, Lim S, Kim JS, et al. (2008) Controlling the thickness of the surface oxide layer on Cu nanoparticles for the fabrication of conductive structures by ink-jet printing. Adv Funct Mater 18: 679–686. doi: 10.1002/adfm.200700902

56. Zhang J, Wang Y, Cheng P, Yao YL (2006) Effect of pulsing parameters on laser ablative cleaning of copper oxides. J Appl Phys 99: 1–11. doi: 10.1063/1.2175467

57. Ghodselahi T, Vesaghi MA, Shafielkhani A, Bachizadeh A, Lameii M (2008) XPS study of the Cu@Cu2O core-shell nanoparticle. Appl Surf Sci 255: 2730–2734. doi: 10.1016/j.apsusc.2008.08.110

58. Dameron CT, Reese RN, Mehra RK, Kortan AR, Carroll PJ, et al. (1989) Biosynthesis of cadmium sulphide quantum semiconductor crystallites. Nature 338: 596–597. doi: 10.1038/338596a0

59. Kowshik M, Vogel W, Urban J, Kulkarni SK, Paknikar KM (2002) Microbial Synthesis of Semiconductor PbS Nanocrystallites. Adv Mater 14: 815–818. doi: 10.1002/1521-4095(20020605)14:11<815::aid-adma815>3.0.co;2-k

60. Kowshik M, Deshmukh N, Vogel W, Urban J, Kulkarni SK, et al. (2002) Microbial synthesis of semiconductor CdS nanoparticles, their characterization, and their use in the fabrication of an ideal diode. Biotechnol Bioeng 78: 583–588. doi: 10.1002/bit.10233

61. Gericke M, Pinches A (2006) Biological synthesis of metal nanoparticles. Hydrometallurgy 83: 132–140. doi: 10.1016/j.hydromet.2006.03.019

62. Gericke M, Pinches A (2006) Microbial production of gold nanoparticles. Gold Bull 39: 22–28. doi: 10.1007/bf03215529

63. Agnihotri M, Joshi S, Kumar R, Zinjardes S (2009) Kulkarnis (2009) Biosynthesis of gold nanoparticles by the tropical marine yeast Yarrowia lipolytica NCIM3589. Mat Lett 63: 1231–1234. doi: 10.1016/j.matlet.2009.02.042

64. Jha AK, Prasad K, Prasad K (2009) A green low-cost biosynthesis of Sb2O3 nanoparticles. Biochem Eng J 43: 303–306. doi: 10.1016/j.bej.2008.10.016

65. Kowshik M, Ashtaputer S, Kharrazi S, Vogel W, Urban J, et al. (2003) Extracellular synthesis of silver nanoparticles by a silver-tolerant yeast strain MKY3. Nanotechnol 14: 95–100. doi: 10.1088/0957-4484/14/1/321

66. Honary S, Barabadi H, Gharaei-Fathabad E, Naghib F (2012) Green synthesis of copper oxide nanoparticles using Penicillium aurantiogriseum, Penicillium citrinum and Penicillium waksmanii. Dig J Nanomater Bios 7: 999–1005. doi: 10.4314/tjpr.v12i1.2

# PART III

# SEMI-CONDUCTOR INDUSTRY

The semiconductor industry consumes an ever-increasing amount of water. In consequence, it also ends up discharging large volumes of wastewater. According to Global Water Intelligence, industry statistics indicate that creating an integrated circuit on a 300 mm wafer requires approximately 2200 gallons of water in total. In addition to the extremely high water requirements of the manufacturing process, toxic materials and chemicals are used in the fabrication process, ranging from arsenic to harsh acids. These become part of the effluent released into the industry's wastewater, creating a dangerous stew that enters our waterways.

# PART III

# SEMI-CONDUCTOR INDUSTRY

# CHAPTER 7

# Application of Ozone Related Processes to Mineralize Tetramethyl Ammonium Hydroxide in Aqueous Solution

CHYOW-SAN CHIOU, KAI-JEN CHUANG, YA-FEN LIN, HUA-WEI CHEN, AND CHIH-MING MA

## 7.1 INTRODUCTION

The semiconductor industry is an important component of the electronics industry, whose global market yield has already exceeded that of the automobile industry. Anisotropic chemical wet etching is widely used in the semiconductor industry to fabricate microstructures on single crystal silicon wafers [1]. Of all the anisotropic etchants, the inorganic KOH (potassium hydroxide) and organic TMAH (tetramethyl ammonium hydroxide) solutions are the most commonly used [1, 2]. Moreover, TMAH solution has also attracted attention because it is clean room compatible, nontoxic, and easy to handle. It also exhibits excellent selectivity to silicon oxide and silicon nitride masks [3, 4]. It has been estimated that, for an 8 in

*Application of Ozone Related Processes to Mineralize Tetramethyl Ammonium Hydroxide in Aqueous Solution.* © 2013 Chyow-San Chiou et al. International Journal of Photoenergy *Volume 2013 (2013), Article ID 191742 (http://dx.doi.org/10.1155/2013/191742). Creative Commons Attribution License 3.0 http://creativecommons.org/licenses/by/3.0/*

wafer manufacturing facility with a monthly production of 20,000 wafer units, TMAH is by far the most concentrated chemical in wastewater [5]. Biological processes are the most widely accepted treatment for organic wastewater of both domestic and industrial origins; however, available information on the biodegradability of TMAH is scarce. Chemical oxidation involving various forms of advanced oxidation processes (AOPs) can be employed as preliminary treatment to convert the potentially biorefractory compounds into intermediate products that are more amenable to biodegradation [6].

Ozone ($O_3$) is a chemical agent widely used for the mineralization (i.e., transformation into $CO_2$ and inorganic ions) of herbicides and related biorecalcitrant organic contaminants in water [7]. Disadvantages of ozonation alone ($O_3$ system) for water treatment are the high energy cost required for its generation and very limited mineralization of refractory COD in industrial effluents. Indeed, hydroxyl radical is a less selective and more powerful oxidant than molecular ozone. A common objective of AOPs is to produce a large amount of radicals (especially –OH) to oxidize the organic matter. Alternative procedures involving ozonation catalyzed with $H_2O_2$ [5], UV light [6], catalysts [1], and $Fe^{2+}$ [8, 9] allow a quicker removal of organic pollutants, because such catalysts improve the oxidizing power of $O_3$ yielding a significant reduction of its economic cost.

The present study assessed the function of UV light, magnetic catalyst ($SiO_2/Fe_3O_4$), and $H_2O_2$ on the enhancing $O_3$ to mineralize TMAH. A concentration of total organic carbon (TOC) was chosen as a mineralization index of decomposition of TMAH. The effects of pH value of aqueous solution and ionic strength, Cl–, on the mineralization of TMAH were examined in this study.

## 7.2 MATERIAL AND METHODS

The batch experiments of mineralization reaction were conducted in a 2.3 L glass flask reactor as illustrated in Figure 1. The UV irradiation source was two 8W lamps encased in a quartz tube with wavelengths of 254 nm. A UVX Radiometer (UVP Inc., USA) was employed for the determination of UV light intensity. The UV intensity of one 8W UV lamp at 254 nm is

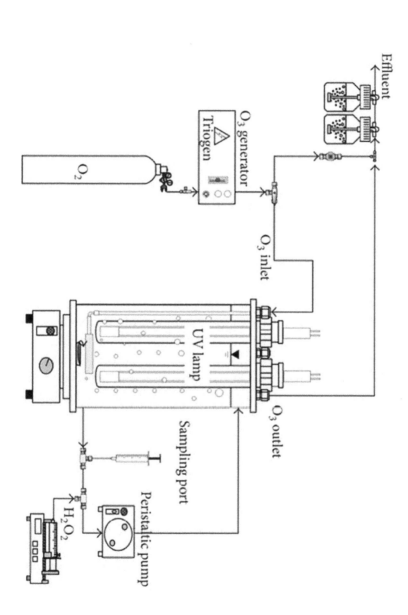

**FIGURE 1:** Experimental sketch of UV/H$_2$O$_2$/O$_3$ system.

$18.6 mW/cm^2$. The ozone generator is from Triogen with the capacity of 10 g/hr. The flow rate of ozone air stream was 4 L/min directed into the photoreactor, and the inlet ozone concentration was 26mg/L.The concentration of ozone was analyzed online by an ozone analyzer (Anseros, Ozamat GM-6000-PRO). $H_2O_2$ was added into the reactor by a syringe pump at constant dosage rate. The pH value of the solution was controlled by the addition of 0.01N $H_2SO_4$/NaOH during the whole reaction time. The effects of pH values, ionic strength, and the initial concentration of Cl⁻ on mineralization efficiency of the $UV/O_3$ process were examined by varying one factor while keeping the other parameters fixed.

### 7.2.1 CATALYSTS PREPARATION

Magnetite ($Fe_3O_4$) was purchased from Sigma-Aldrich (St. Louis, MO, USA) and used without any further purification. A total of 1.08 L of aqueous solution containing 20 g of $Fe_3O_4$ particles was held in a 2 L beaker at 90°C; the pH was maintained at 9.5 with 0.1N NaOH, while being stirred by a mechanic stirrer. An appropriate amount of 20 g $Na_2O \cdot nSiO_2$ was dissolved in 100mL of deionized water; the aqueous solution was then mixed with the aqueous solution containing magnetite (Fe3O4) with a mechanic stirrer for 30 min. At last, the magnetic catalysts (i.e., $SiO_2/Fe_3O_4$) were dried at 105°C, after the pH value of the slurry solution was maintained at 8 with 5N $H_2SO_4$.

### 7.2.2 INSTRUMENTAL ANALYSIS

Tetramethyl ammonium hydroxide, TMAH ($C_4H_{10}NO$, MW = 98), supplied by Aldrich (USA) was used as the target compound in this study. Hydrogen peroxide ($H_2O_2$) of 35wt.% supplied by Shimakyu Co. (Japan) was injected into the reactor at a constant feed rate by syringe pump. All chemicals from several suppliers were reagent grade. The mineralization efficiency of TMAH by this advanced oxidation process was determined by the analytical results of a TOC analyzer (Tekmar, Dohrmann Phoenix 8000). $Na_2SO_3$ solution (1.0 g/L) and Spectroquant Picco colorimeter test kit (Merck, Germany) was used to quench and measure residual dissolved ozone in samples for TOC analysis, respectively. This instrument utilizes the UV persulfate technique to convert organic carbon into carbon dioxide ($CO_2$), analyzed by an infrared $CO_2$ analyzer and calibrated with potassium hydrogen phthalate.The magnetic properties and the isoelectric point (IEP) of the catalysts were determined by a vibrating sample magnetometer (Lake Shore, 7407) and a Zetasizer (Nano ZS ZEN 3600), respectively.

## 7.3 RESULTS AND DISCUSSION

### 7.3.1 SURFACE CHARACTERISTICS OF MAGNETIC PHOTOCATALYST

Figure 2 showed the surface zeta potential versus pH for the magnetic catalyst $SiO_2/Fe_3O_4$, and this same process was repeated two times.The pH of the isoelectric point (IEP) for the magnetic catalyst $SiO_2/Fe_3O_4$ was 2.8. The IEP for $SiO_2$ particles and $Fe_3O_4$ particles was previously determined by other procedures and ranged from 2 to 3 and from 6.5 to 6.8, respectively [10]. Therefore, the results observed that $Fe_3O_4$ (core) was almost covered by $SiO_2$ (shell) because the IEP of $SiO_2/Fe_3O_4$ was close to $SiO_2$ particles.

The magnetic properties of the $SiO_2/Fe_3O_4$ and $Fe_3O_4$ core were measured with a vibrating sample magnetometer (VSM), as shown in Figure 3. The M-H plots showed the change inMs of the particles, after the incorporation of a $SiO_2$ shell. The saturation magnetization (Ms) was 39.2 emu $g^{-1}$ and was observed in $SiO_2/Fe_3O_4$. The results indicated that the prepared samples exhibited paramagnetic behaviors at room temperature [11].

### 7.3.2 MINERALIZATION EFFICIENCY OF TMAH UNDER VARIOUS CONDITIONS

To confirm the roles of $O_3$, UV, and $H_2O_2$ in the mineralization reaction of TMAH, five sets of experiments were performed to compare the mineralization efficiency of TMAH under various conditions as a function of

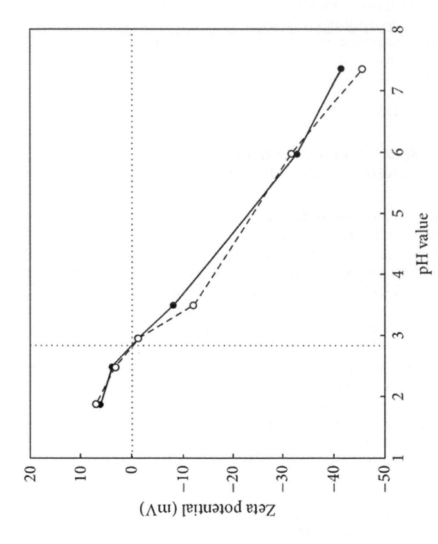

**FIGURE 2:** Zeta potential of magnetic catalysts ($SiO_2/Fe_3O_4$).

time, which is given as $\eta_{TOC,TMAH} = (TOC_0 - TOC)/TOC_0$, and the results are shown in Figure 4. Condition (a) denotes a reaction system with UV 254 nm, power density of 37.2mW/cm², and $dC_{H2O2}/dt$ of 2.5 × 10⁻⁴ mol/min (called UV/H₂O₂). The obtained $\eta_{TOC,TMAH}$ was 38.5% after 60 min of reaction time. We also attempted to mineralize TMAH by H₂O₂ alone, and the result indicated that H₂O₂ does not possess the ability to mineralize TMAH. As is the case with UV, with H₂O₂, theoretically, the photolysis of one mole H₂O₂ leads to two moles –OH according to (1), and the resulting –OH with higher oxidation potential could mineralize TMAH (38.5%) as follows:

$$H_2O_2 + h\upsilon \rightarrow 2 \cdot OH \tag{1}$$

Condition (b) represents a reaction system with only O₃, and the resulting percentage of $\eta_{TOC,TMAH}$ was approximately 49.2% after 60min of reaction time. There are two mechanisms through which O₃ can degrade organic pollutants, namely, (i) direct attack and (ii) indirect attack through the formation of hydroxyl radicals [12]. The observed $\eta_{TOC,TMAH}$ is better than that for UV/H₂O₂. Under condition (c), using H₂O₂ and O₃ (denoted here as H₂O₂/O₃), a better $\eta_{TOC,TMAH}$ (approximately 58.1% at $t$ = 90min) was achieved as compared to the values under conditions (a) and (b), thereby confirming the strong oxidation ability of H₂O₂/O3. The oxidation potential of H₂O₂/O3 is based on the fact that the conjugate base of H₂O₂ can catalyze ozone into the formation of ·OH(Gottschalk et al., 2000). Anoptimumdose ratio of H₂O₂/O₃ has often been shown to be in the molar range of 0.5–1 depending on the presence of promoters and scavengers. H₂O₂ itself can act as a scavenger as well as an initiator, and therefore determining the optimum dose ratio of H₂O₂/O₃ is important [7]. With the stoichiometric molar ratio of H₂O₂/O₃ being 0.5, the overall reaction of H₂O₂ catalyzes O₃ to produce ·OH as shown in (2). In this study, the dosage rates of O₃ and H₂O₂ are 5.0 × 10⁻⁴ mol/min and 2.5 × 10⁻⁴ mol/min, respectively; consider the following

$$H_2O_2 + 2O_3 \rightarrow 2 \; OII + 3O_2 \tag{2}$$

Condition (d), in which UV and $O_3$ (denoted as UV/$O_3$) were employed, yielded a value of approximately 87.6% for $\eta_{TOC,TMAH}$ after 60 min of reaction time, which is far better than conditions (a), (b), and (c). Due to the strong photolysis of ozone in combination with UV radiation ($\varepsilon$ 254nm = 3300$M^{-1}cm^{-1}$ for $O_3$), the decay rate of ozone for decay resulting from the UV is higher than that resulting from $H_2O_2$ by a factor of approximately 1000 [7]; this results in the production of more hydroxyl radicals according to (3) and (4), which in turn results in higher mineralization efficiency, as follows:

$$O_3 + h\upsilon \rightarrow O(^1D) + O_2^* \qquad\qquad (3)$$

$$O(^1D) + H_2O \rightarrow 2 \cdot OH \qquad\qquad (4)$$

Condition (e), involving UV, $O_3$, and $H_2O_2$ (denoted as UV/$H_2O_2$/$O_3$), resulted in a poor efficiency ($\eta_{TOC,TMAH}$ = 72.3% at $t$ = 60 min) as compared to condition (d). In the UV/$O_3$ process, the amount of residual $O_3$ in the aqueous solution was small due to the high decay rate of $O_3$ by UV irradiation, and subsequently $H_2O_2$, when added, will react with $\cdot$OH instead of with $O_3$ as indicated in (5). Under this condition, $H_2O_2$ behaves as a scavenger of $\cdot$OH and a small value of $\eta_{TOC,TMAH}$ results. In this study, we also evaluated the effect of the addition of $H_2O_2$ on the mineralization efficiency of UV/$O_3$, and the experimental results (data not shown) indicated that even the relatively low $H_2O_2$ dosage of $dC_{H2O2}/dt$ = 5.0 × $10^{-5}$ mol/min causes the molar ratio of $H_2O_2$/$O_3$ to have a value considerably below 0.5. The mineralization efficiency of UV/$O_3$ with the addition of $H_2O_2$ is poorer than that without the addition of $H_2O_2$. This result led to the conclusion that adding $H_2O_2$ into the UV/$O_3$ processwill result in a negative effect on $\eta_{TOC,TMAH}$ as follows:

$$H_2O_2 + \cdot OH \rightarrow H_2O + \cdot O_2H \qquad\qquad (5)$$

Condition (f), involving $SiO_2$/$Fe_3O_4$, $O_3$, and $H_2O_2$ (denoted as $SiO_2$/$Fe_3O_4$/$H_2O_2$/$O_3$), resulted in a high efficiency ($\eta_{TOC,TMAH}$ = 69.0% at $t$ = 60 min) as compared to condition (a). To evaluate the ability of catalyst adsorption, the adsorption experiment at $SiO_2$/$Fe_3O_4$ = 0.2g and initial

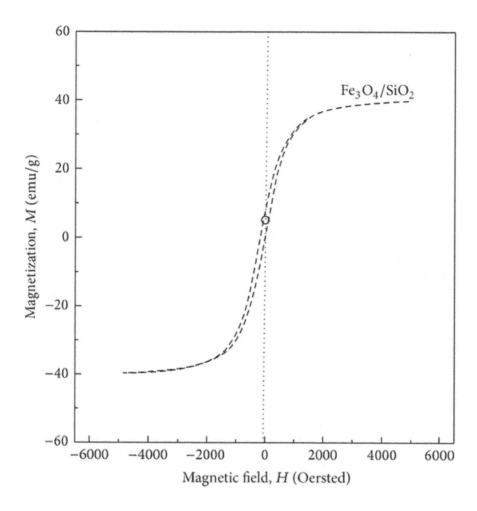

**FIGURE 3:** Magnetization versus applied magnetic field formagnetic catalysts (SiO$_2$/Fe$_3$O$_4$).

concentration of TMAH = 40mg/L revealed that the adsorption of TMAH was less than 10% within 60 minutes. The mineralization efficiency of SiO$_2$/Fe$_3$O$_4$/H$_2$O$_2$/O$_3$ was significantly higher than that of pure O$_3$, possibly because SiO$_2$/Fe$_3$O$_4$ and H$_2$O$_2$ could react with the dissolved O$_3$ molecules to generate reactive oxidative species ($O_2^-$, $HO_2 \cdot$, $\cdot OH$, and $O_3^-$) via free-radical chain reactions. The initiator ($H_2O_2$) could induce the formation of superoxide ions ($O_2^-$) from $O_3$ molecules, and $\cdot OH$ formed in the chain reaction was used for the mineralization of organic compounds [13]. Furthermore, the paramagnetic behaviors of the prepared $SiO_2/Fe_3O_4$ gave rise to the magnetic catalyst $SiO_2/Fe_3O_4$, which could be separated more easily through the application of a magnetic field. According to the experimental results, >90% of the magnetic catalyst was recovered and easily redispersed in a solution for reuse.

Condition (h), involving $SiO_2/Fe_3O_4$ and $O_3$ (denoted as $SiO_2/Fe_3O_4/O_3$), resulted in a high efficiency ($\eta_{TOC,TMAH} = 60.1\%$ at $t = 60min$) as compared to condition (a). The mineralization efficiency of $SiO_2/Fe_3O_4/O_3$ was significantly higher than that of pure $O_3$ possibly because of the enhancement of reactive oxidative species with the chain reaction of $SiO_2/Fe_3O_4$ and the dissolved $O_3$.

As a result, the mineralization efficiency of TMAH under various conditions follows the sequence: $UV/O_3$ > $UV/H_2O_2/O_3$ > $H_2O_2/SiO_2/Fe_3O_4/O_3$ > $H_2O_2/O_3$ > $SiO_2/Fe_3O_4/O_3$ > $O_3$ > $UV/H_2O_2$. Figure 4 presents the variations of TOC and pH of the TMAH solution under the $UV/O_3$ process as a function of time. As shown in Figure 4, the removal of TOC is close to 87.6% at $t = 60min$, indicating that the $UV/O_3$ process could mineralize TMAH efficiently. Furthermore, the pHvalue of the reaction solution without a buffer system decreased considerably from 10 to 4.5 during the entire reaction time, thereby revealing that acid intermediates are formed before TMAH is converted into $CO_2$.

## 7.3.3 EFFECT OF PH ON THE MINERALIZATION EFFICIENCY OF UV/O3

The direct attack on organic pollutants by molecular ozone (commonly known as ozonolysis) occurs under acidic or neutral conditions. At a high pH value, ozone decomposes to nonselective hydroxyl radicals according to (6), which in turn oxidizes the organic pollutants. Many researches [6, 12] found that an increasing pH accelerates ozone decomposition to generate hydroxyl radicals, which destroy organic compounds more effectively

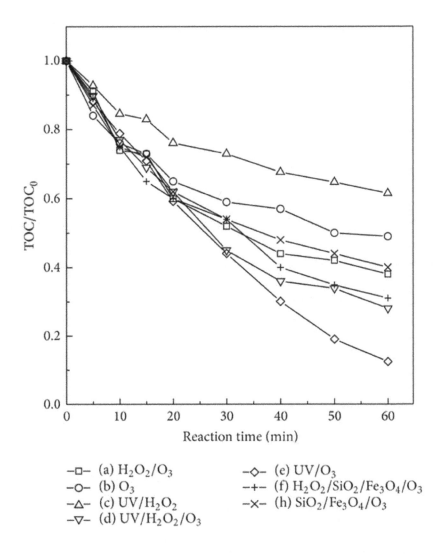

−□− (a) $H_2O_2/O_3$        −◇− (e) $UV/O_3$
−○− (b) $O_3$              −+− (f) $H_2O_2/SiO_2/Fe_3O_4/O_3$
−△− (c) $UV/H_2O_2$         −×− (h) $SiO_2/Fe_3O_4/O_3$
−▽− (d) $UV/H_2O_2/O_3$

**FIGURE 4:** Dependence of mineralization of TMAH on time at various conditions. Experimental conditions: case (a): UV ($\lambda_{254}$) = 37.2mWcm$^{-2}$, dosing rate of $H_2O_2$ (d/C$_{H2O2}$/dt) = 2.5×10$^{-4}$ mol/min; case (b): dosing rate of $O_3$ (d/C$_{O3}$/dt) = 5.0×10$^{-4}$ mol/min; case (c): d/C$_{H2O2}$/dt = 2.5×10$^{-4}$ mol/min, d/C$_{O3}$/dt = 5.0×10$^{-4}$ mol/min; case (d): UV ($\lambda_{254}$) = 37.2mWcm$^{-2}$, d/C$_{O3}$/dt = 5.0×10$^{-4}$ mol/min; case (e): UV ($\lambda_{254}$) = 37.2mWcm$^{-2}$, d/C$_{O3}$/dt = 5.0×10$^{-4}$ mol/min, d/C$_{H2O2}$/dt = 2.5×10$^{-4}$ mol/min. The initial concentration of TMAH ($C_{TMAH,0}$) for all cases was 40mg/L; case (f): SiO$_2$/Fe$_3$O$_4$ = 0.2 g, d/C$_{O3}$/dt = 5.0×10$^{-4}$ mol/min, d/C$_{H2O2}$/dt = 2.5×10$^{-4}$ mol/min. The initial concentration of TMAH ($C_{TMAH,0}$) for all caseswas 40mg/L.

than ozone. Therefore, the pH of aqueous solution is an important factor that determines the efficiency of ozonation since it can alter the degradation pathways as well as kinetics. One has

$$O_3 + OH^- \rightarrow \cdot OH^+ (\cdot O_2 \longleftrightarrow \cdot O_2H) \qquad\qquad (6)$$

For the combined oxidation process, $UV/O_3$, the effect of pH on the mineralization efficiency is more complex. The study [14] found that neither low pH values nor high pH values of the $UV/O_3$ could provide a degradation rate better than that obtained by the simultaneous application of $UV/O_3$ with neutral pH values.

The pH values of the aqueous solution were controlled to stay between 3 and 10 to evaluate the pH effect on the mineralization efficiency of TMAH by the $UV/O_3$ process, as shown in Figure 5. As the pseudo-first-order kinetic hypothesized, Table 1 reveals that the influence of the pH value on the reaction rate is negligible at pH values in the range from 3 to 10. It is clear from (6) that more hydroxyl radicals were produced at high pH values, thus enhancing the mineralization rate of TMAH. However, the production of hydroxyl radicals by the $UV/O_3$ process also proceeds according to (3) and (4), and it was not influenced by the pH value of the aqueous solution. Furthermore, the high $OH^-$ ion content of the system

**TABLE 1:** The pseudo-first-order rate constant $K_{obs}$, half-life $t_{1/2}$, and correlation coefficients for degradation of TMAH by $UV/O_3$ at different pH values. (Experimental conditions: pH values were adjusted by $H_2SO_4$ and NaOH and fixed at a constant value during the whole reaction time; the other conditions were the same as in Figure 4(d).)

| pH | $K_{obs}$ (1/min) | $R^2$ |
|----|-------------------|-------|
| 3.0 | 0.0315 | 0.942 |
| 5.0 | 0.0313 | 0.917 |
| 7.5 | 0.0312 | 0.973 |
| 10.0 | 0.0318 | 0.907 |

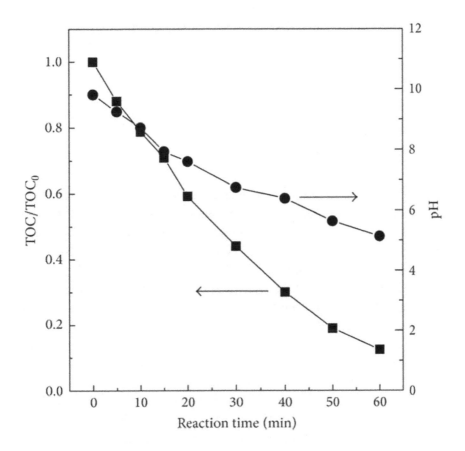

**FIGURE 5:** Time variation of TOC and pH using   process to mineralize TMAH. Experimental conditions were the same as those of case (d) in Figure 4.

may trap the mineralization generated $CO_2$ in the solution and, as a result, bicarbonates and carbonates are formed in the alkaline system. Both bicarbonates and carbonates are efficient scavengers of hydroxyl radicals due to their very high reaction rate constants with the hydroxyl radicals ($k = 8.5 \times 10\ 6M^{-1}\ s^{-1}$ for bicarbonates and $k = 3.9 \times 10\ 8M^{-1}\ s^{-1}$ for carbonates).Thus, due to the influence of the increase in hydroxyl radicals and the formation of scavengers, the comprised results causing the pH effect on the mineralization of TMAH by $UV/O_3$ are negligible for pH values of the solution in the range from 3 to 10. Note that a buffer system was not introduced in the later experiments, except for the experiment relating to the pH effect.

## 7.3.4 EFFECT OF CHLORIDE ION AND IONIC STRENGTH ON THE MINERALIZATION

The effect of chloride ions, which are frequently present in industrial wastewater, on the mineralization efficiency of TMAH with $UV/O_3$ was evaluated, as shown in Table 2. The mineralization rate of the TMAH solution containing chloride ion by the $UV/O_3$ process could not be expressed by the pseudo-first-order kinetic. So, the mineralization efficiency shown in Table 2 was illustrated by $\eta_{TOC,TMAH}$ at the reaction time of 60min. The experimental results indicate that $\eta_{TOC,TMAH}$ decreased with an increase in the chloride ion concentration. Chloride ions are likely to retard the efficiency of the mineralization of TMAH by competing for the oxidizing hydroxyl radicals and ozone molecules. Chloride ions can be oxidized by ozone as per (7) and (8) [15], and they can be converted into $ClO^-$ and $Cl_2$. Thus, the effective concentration of ozone was decreased by chloride ions, and the oxidation potential of the resulting products, HOCl and $Cl_2$, was lower than that of ozone. Furthermore, chloride ions may also act as scavenger with regard to the hydroxyl radical as per (9) [15]. One has

$$O_3 + Cl^- + H^+ \rightarrow HOCl + O_2 \qquad (7)$$

$$O_3 + 2Cl^- + 2H^+ \rightarrow Cl_2 + H_2O + O_2 \qquad (8)$$

$$Cl^- + \cdot OH \rightarrow HOCl^{\cdot -} \qquad (9)$$

Ionic strength may affect the effective concentration of a compound in a solution and this becomesmore significant in the presence of polar compounds [16]. As noted, TMAH is a quaternary ammonium compound. Its ammonium ions are surrounded by anions in solutions, resulting in

**TABLE 2:** The mineralization efficiency of TMAH by UV/O$_3$ at different chloride concentrations. (Experimental conditions are as shown in Figure 4(d).)

| Cl$^-$ (mg/L) | $\eta_{TOC, TMAH}$ (%) |
|---|---|
| 0 | 87.56 |
| 100 | 80.08 |
| 200 | 78.58 |
| 300 | 76.05 |
| 500 | 74.73 |

shielding off oxidizing agents such as O$_3$ and hydroxyl radicals. Presumably, it decreases the mineralization efficiency of TMAH by the UV/O$_3$ process. The experimental study [16] examined the effect of ionic strength on the solubility of O$_3$ for various types of inorganic solutions. These researchers concluded that there is no significant effect on the solubility of O$_3$ in sulfate solutions. Additionally, sulfates are not oxidized by ozone molecules and hydroxyl radicals. In the present study, attempts have been made to evaluate the effect of ionic strength on the mineralization of TMAH in sulfate solutions by the UV/O$_3$ process, and the results are shown in Table 3. As can be seen in the table, the variation of the ionic strength neither facilitates nor suppresses the mineralization of TMAH. This indicates that, in the UV/O$_3$ process, the inhibition of the mineralization of polar organic compounds in an aqueous solution is not significant at high ionic strengths.

The reuse experiments were carried out by evaluating the stability of catalyst activity. In this experiment, 0.2 g L$^{-1}$ of magnetic catalyst (SiO$_2$/Fe$_3$O$_4$) was used at an initial concentration of 40mg/L of TMAH. After the ozonation process, the magnetic catalyst (SiO$_2$/Fe$_3$O$_4$) was collected by magnetic force. The clear solution was used for analytical determination, and the magnetic catalyst was used directly in the subsequent catalytic ozonation process. This same process was repeated four times and the removal of TMAH in the reuse experiment is shown in Figure 6. The catalytic activity of $SiO_2/Fe_3O_4$ remained constant and no obvious deactivation (<15%) was observed after being used four times. From the results

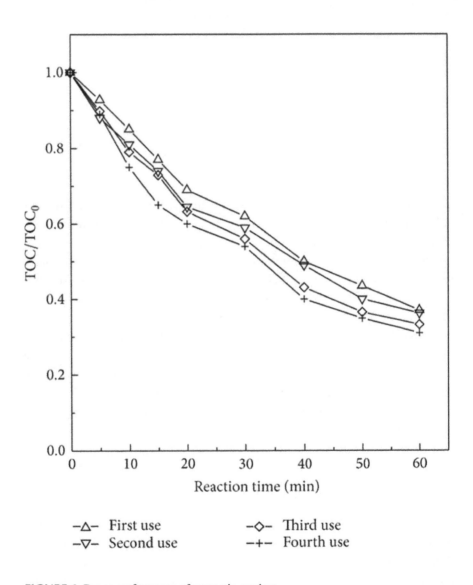

**FIGURE 6:** Reuse performance of magnetic catalyst.

**TABLE 3:** The pseudo-first-order rate constant $K_{obs}$, half-life $t_{1/2}$, and correlation coefficients for degradation of TMAH by $UV/O_3$ at different concentrations of $K_2SO_4$. (The other experimental conditions were the same as in Figure 4(d).)

| K2SO4 (mg/L) | $K_{obs}$ (1/min) | $R^2$ |
|:---:|:---:|:---:|
| 0 | 0.0320 | 0.977 |
| 100 | 0.0309 | 0.948 |
| 300 | 0.0313 | 0.969 |
| 500 | 0.0307 | 0.981 |

of the reuse and recovery experiments, the magnetic catalyst $SiO_2/Fe_3O_4$ is considered to show considerable promise in water treatment use.

## 7.4 CONCLUSIONS

The major results of applying the advanced oxidation process to mineralize TMAH can be summarized as follows.

(1) The rank of treatment conditions based on the mineralization efficiency of TMAH has the sequence: $UV/O_3 > UV/H_2O_2/O_3 > H_2O_2/SiO_2/Fe_3O_4/O_3 > H_2O_2/O_3 > SiO_2/Fe_3O_4/O_3 > O_3 > UV/H_2O_2$.

(2) The experimental results of this study suggest that UV irradiation 37.2mWcm−2 UV (254 nm) and $O_3$ flow rate of $0.5 \times 10{-4}$ mol/min provide the best condition for the mineralization of TMAH (20mg/L), resulting in 95% mineralization, at 60 min reaction time. Adding $H_2O_2$ into the UV/O3 process will suppress the mineralization efficiency.

(3) The mineralization efficiency of $SiO_2/Fe_3O_4/H_2O_2/O_3$ was significantly higher than that of $O_3$, $H_2O_2/O_3$ $UV/H_2O_2$. More than 90% of the magnetic catalyst was recovered and easily redispersed in a solution for reuse. The addition of chloride ions in reaction solution will suppress the mineralization efficiency of the $UV/O_3$ process. Ionic strength and variable pH values (from 3 to 10) in reaction solution show no effect on the mineralization efficiency of the $UV/O_3$ process.

## REFERENCES

1.  H. W. Chen, Y. Ku, and Y. L. Kuo, "Photodegradation of o-Cresol with Ag deposited on TiO2 under visible and UV light irradiation," Chemical Engineering and Technology, vol. 30, no. 9, pp. 1242–1247, 2007.

2.  M.-L. Wang and B.-L. Liu, "Kinetics of two-phase reaction of o-phenylene diamine and carbon disulfide catalyzed by tetrabutylammonium hydroxide in the presence of potassium hydroxide," Journal of the Chinese Institute of Engineers, vol. 30, no. 3, pp. 423–430, 2007.

3.  P. M. Sarro, D. Brida, W. V. D. Vlist, and S. Brida, "Effect of surfactant on surface quality of silicon microstructures etched in saturated TMAHW solutions," Sensors and Actuators A, vol. 85, no. 1, pp. 340–345, 2000.

4.  I. Zubel and M. Kramkowska, "The effect of isopropyl alcohol on etching rate and roughness of (100) Si surface etched in KOH and TMAH solutions," Sensors and Actuators A, vol. 93, no. 2, pp. 138–147, 2001.

5.  W. Den, F.-H. Ko, and T.-Y. Huang, "Treatment of organic wastewater discharged from semiconductor manufacturing process by ultraviolet/hydrogen peroxide and biodegradation," IEEE Transactions on Semiconductor Manufacturing, vol. 15, no. 4, pp. 540–551, 2002.

6.  S.-P. Tong, D.-M. Xie, H. Wei, and W.-P. Liu, "Degradation of sulfosalicylic acid by O3/UV O3/TiO2/UV, and O3/V-O/TiO2: a comparative study," Ozone: Science and Engineering, vol. 27, no. 3, pp. 233–238, 2005.

7.  C. Gottschalk, J. A. Libra, and A. Saupe, Ozonation of Water and Waste Water, Wiley-VCH, New York, NY, USA, 2000.

8.  T. K. Chen, C. H. Ni, and J. N. Chen, "Nitrification-denitrification of opto-electronic industrial wastewater by anoxic/aerobic process," Journal of Environmental Science and Health A, vol. 38, no. 10, pp. 2157–2167, 2003.

9.  P. Cañizares, R. Paz, C. Sáez, and M. A. Rodrigo, "Costs of the electrochemical oxidation of wastewaters: a comparison with ozonation and Fenton oxidation processes," Journal of Environmental Management, vol. 90, no. 1, pp. 410–420, 2009.

10. D. E. Keller, D. C. Koningsberger, and B. M. Weckhuysen, "Elucidation of the molecular structure of hydrated vanadium oxide species by X-ray absorption spectroscopy: correlation between the V…V coordination number and distance and the point of zero charge of the support oxide," Physical Chemistry Chemical Physics, vol. 8, no. 41, pp. 4814–4824, 2006.

11. J. Xu, Y. Ao, D. Fu, and C. Yuan, "Low-temperature preparation of anatase titania-coated magnetite," Journal of Physics and Chemistry of Solids, vol. 69, no. 8, pp. 1980–1984, 2008.

12. J. Hoigne and H. Bader, "The role of hydroxyl radical reactions in ozonation processes in aqueous solutions," Water Research, vol. 10, no. 5, pp. 377–386, 1976.

13. B. Kasprzyk-Hordern, M. Ziółek, and J. Nawrocki, "Catalytic ozonation and methods of enhancing molecular ozone reactions in water treatment," Applied Catalysis B, vol. 46, no. 4, pp. 639–669, 2003.

14. T. S. Müller, Z. Sun, M. P. Gireesh Kumar, K. Itoh, and M. Murabayashi, "The combination of photocatalysis and ozonolysis as a new approach for cleaning

2,4-dichlorophenoxyaceticacid polluted water," Chemosphere, vol. 36, no. 9, pp. 2043–2055, 1998.

15. J. de Laat and T. G. Le, "Effects of chloride ions on the iron(III)-catalyzed decomposition of hydrogen peroxide and on the efficiency of the Fenton-like oxidation process," Applied Catalysis B, vol. 66, no. 1-2, pp. 137–146, 2006.

16. D. A. Skoog, D. M. West, F. J. Holler, and S. R. Crouch, Analytical Chemistry, Thomson Learning, Singapore, 2000.

# CHAPTER 8

# Physico-chemical Investigation of Semiconductor Industrial Wastewater

Y. C. WONG, V. MOGANARAGI, AND N. A. ATIQAH

## 8.1 INTRODUCTION

Precipitation, surface water runoff, surface water and groundwater storage and evaporation are involved in the hydrologic cycle, which caused the occurrence of changes to the quality of the water body. For instance, precipitation in the form of rain or snow can bring airborne pollutants to the Earth's surface; surface water runoff can cause erosion andtransport sediments; groundwater recharge can leach chemicals into aquifers; and evaporation can elevate concentrations of pollutants in bodies of water by dropping the total volume of stored water. Each natural component of the hydrologic cycle can have a negative effect on surface and groundwater quality. As human beings, we also have great effects on water quality. We add waste to the environment through the utilization of resources such as food, clothing, housing, and fuel for transportation. The fast growing world population is contributing to the fall of our existing water quality and is creating significant challenges for water managers, industry and fish and wildlife agencies [1].

*Physico-chemical Investigation of Semiconductor Industrial Wastewater.* © *Wong Y. C, Moganaragi V, Atiqah N. A.* Orient J Chem *2013;29(4) (http://www.orientjchem.org/?p=1314). Used with permission of the* Oriental Journal of Chemistry.

The water quality measurement consists of several parameters that can be measured using proper methods. Some simple measurements such as temperature, pH, dissolved oxygen, conductivity, salinity and total dissolved solid can be measured on the site [2].

Wastewater quality can be defined by physical, chemical, and biological characteristics. Physical parameters include colour, odour, temperature, solids (residues), turbidity, oil, and grease. Chemical parameters associated with the organic content of waste water include the biochemical oxygen demand (BOD), chemical oxygen demand (COD), total organic carbon (TOC), and total oxygen demand (TOD). Inorganic chemical parameters include salinity, hardness, pH, acidity, alkalinity, iron, manganese, chlorides, sulfates, sulfides, heavy metals (lead, chromium, copper, and zinc), nitrogen (organic, ammonia, nitrite, and nitrate), and phosphorus whereas the bacteriological parameters include coliforms, fecal coliforms, specific pathogens, and viruses [3-4].

The most general physical characteristics of water are colour, odour, temperature, taste, and turbidity while biological characteristics of water are living organisms including bacteria (e.g. Escherichia coli), viruses, protozoans (e.g. Cryptosporidiosis), phytoplankton (i.e. microscopic algae), zooplankton (i.e. tiny animals), insects, plants, and fish [5].

For the past few years, electronic industries are well developed sector in many countries around the world. In coming years, this electronic industry is predicted to expend its growth to a wider prospect. Semiconductors are one of the essential products which are produced by the electronic industries. The semiconductors are used in various types of equipment such as computer devices, telecommunication devices, consumer electronic products, electronic control devices, scientific and also in medical test equipments. The processes involved in the production of semiconductors consisted of complex and highly mere processes including silicon growth, oxidation, doping, photolithography, etching, stripping, dicing, metallization, planarization, cleaning, and etc. [6,7,8]. The manufacturing processes of semiconductor integrated circuits involved over two hundred types of organic and inorganic substances (proprietary and generic). Some of the steps in the sequences are water back grinding, sawing, die attach, wire bonding, encapsulation, electroplating, trim and form and marking[8-9].

HACH Company (2010) showed that the colour of the semiconductor wastewater was milky and this was because of the occurrence of fine suspended oxide particles. The total solid (TS) contents in semiconductor waste water was high, which was 4500 mg/L but the suspended solids (SS) concentrations of both samples were small, being less than 0.2 mg/L. This was as a result of the fact that the size of the fine suspended oxide particles was below the pore size 0.45 μm of the standard filter paper used for the SS measurements while the TS concentration was calculated by drying the sample in a crucible which retained all the fine oxide particles [10].

In addition,the semiconductor wastewater used in their study had a pH 9.5 and very low suspended solids concentration at 10 mg/L, and high chemical oxygen demand (COD) concentration of over 70000 mg/L. They also found that the biodegradability of the wastewater as represented by the ratio of BOD/COD was low at 0.124, reflecting the existence of recalcitrant organic compounds. They further noted that the COD concentration of the semiconductor wastewater was due to mixing of the semiconductor wastewater with other processes wastewater streams that contained organic compounds, but little total solids. Hence, COD removal from the semiconductor wastewater is as important as the removal of fine suspended oxide particles if the semiconductor wastewater is to be upgraded to a level for general reusepurpose [11].

The objectives of this study are to identify the characteristics of industrial wastewater before and after treatment, the quality that must be maintained in the environment to which the wastewater is to be discharged or reuse and provide understanding in the methods involved to analyze various physical and chemical parameters of the wastewater.

## 8.2 METHODOLOGY

### 8.2.1 SAMPLING METHOD

Sample collection is important in deriving relevant data that may be utilized to make important decisions [12]. Samples were collected from a semiconductor company for analysis of various physical and chemical parameters such as temperature, pH, turbidity, total suspended solid (TSS), conductivity, biochemical oxygen demand (BOD), chemical oxygen demand (COD), dissolved oxygen, salinity, chlorine dioxide and heavy metal contents. Samples were taken from two points namely as untreated effluents and also treated effluents and were collected in clean dry plastic bottles in such a way that no bubbles were formed in the bottles. The bottles were foiled with aluminium foil and filled with preservatives such as 3.0ml of sulphuric acid (50%) for COD while 1.5ml of nitric acid (50%) for heavy metal to keep the samples below pH 2 except BOD which did not require any preservatives. Aluminium foil was used to avoid any undesirable chemical reaction by sunlight. After that, the samples were incubated at 4°C prior to any chemical analysis in laboratory. Take note that all of the sample bottles were labelled properly to avoid any confusion.

## 8.2.2 ANALYTICAL PROCEDURES

### 8.2.2.1 DETERMINATION OF TEMPERATURE, PH, CONDUCTIVITY, DISSOLVED OXYGEN AND SALINITY.

Temperature, pH, conductivity, dissolved oxygen and salinity were determined by using a Multi parameter which was measured at the site of the collection. The probe module was rugged, with the sensors enclosed in a heavy duty probe sensor guard with attached sinking weight. A 4, 10 or 20 meter cable was directly connected to the probe module body making it waterproof [13].

### 8.2.2.2 DETERMINATION OF TURBIDITY

Turbidity was determined by using Turbidimeter which was measured at the site of the collection. The HACH 2100P Portable Turbidimetermeasures turbidity from 0.01 to 1000 NTU in automatic range mode with automatic decimal point placement. The manual range mode measures turbidity in three ranges: 0.01 to 9.99, 10 to 99.9 and 100 to 1000 NTU. Designed primarily for field use, the microprocessor-based Model 2100P have the range, accuracy, and resolution of many laboratory instruments [14].

## 8.2.2.3 DETERMINATION OF TOTAL SUSPENDED SOLID (TSS)

TSS was measured by using filtration method. To determine total suspended solids, a piece of filter paper was weighed out as accurately as possible. Then, the water sample was allowed to pass through the conical flask where the filter water was placed in the conical flask. The filter paper was allowed to dry completely and was put in incubator for 24 hours. The change in the weight was the weight of the total suspended solids.

## 8.2.2.4 DETERMINATION OF BIOCHEMICAL OXYGEN DEMAND (BOD)

BOD was measured using HACH HQ 40D (HACH, 2010). The instrument was used with digital IntelliCAL™ probes to measure various parameters in water. The meter recognizes the type of probe that was connected to the meter (HACH, 2009). BOD was measured by first preparing sample dilution water using a BOD Nutrient Buffer Pillow. Then, the mixture was stirred using HACH HQ 40D to remove all the bubbles. After all the bubbles were removed, the reading of the dissolved oxygen content was taken using HACH HQ 40D. Then the sample bottle was wrapped with aluminium foil and incubated for five days.

After five days, the readings for the dissolved oxygen contents after incubation were taken. The readings indicate the amount of dissolved oxygen. BOD measured the rate of oxygen uptake by micro-organisms in a sample of water at a temperature of 20°C and over an elapsed period of five days in the dark. To calculate the BOD, equation 2.1 below was used:-

$BOD = D_1 - D_2 P$     (2.1)

Where, $D_1$ = Dissolved oxygen content before incubation.
$D_2$ = Dissolved oxygen content after incubation after 5 days.
P = the volume of the sample 100 total volume of the bottle 300ml

## 8.2.2.5 DETERMINATION OF CHEMICAL OXYGEN DEMAND (COD)

The DR 5000 Spectrophotometer is a scanning UV/VIS spectrophotom-eter with a wavelength range of 190 to 1100 nm. The DR 5000 is used for testing in visible and ultraviolet wavelengths. A gas filled tungsten lamp produces light in the visible spectrum (320 to 1100 nm), and a deuterium lamp produces light in the ultraviolet spectrum (190 to 360 nm). The DR 5000 Spectrophotometer provides digital readouts in direct concentration units, absorbance, or percent transmittance. When a user generated or pro-grammed method is selected, the on-screen menus and prompts direct the user through the test. Running an analysis with the DR 5000 was rela-tively simple and involved reading of the sample where the instrument measures the amount of light passing through a reacted sample and con-verts the transmitted light into a concentration [15]. COD was measured by homogenizing 100 ml of sample using a blender for 30 seconds. The homogenized sample was stirred with a magnetic stirrer plate. Then, 2.0 ml of homogenized sample was poured into COD buffer, vial, and closed with a cap. The solution was inverted for several times before placed in preheated reactor. The vial was heated for two hours. After two hours, the reactor was turned off and the vial was left to cool to 120°C or less for 20 minutes. Then, it was inverted again several times and placed into a rack to cool to room temperature before it was placed in the cell holder of the spectrophotometer for measurement of COD value. The steps were then repeated with the treated effluents. Chlorine dioxide was also measured using HACH DR 5000 Spectrophotometer. A small amount of sample (10mL) was taken and poured into the sample cell before placing it inside the spectrophotometer. The results were recorded directly from the meter and were repeated for treated effluents.

## 8.2.2.6 DETERMINATION OF HEAVY METALS

For the analysis of heavy metals which were Copper (Cu), Zinc (Zn), Iron (Fe), Manganese (Mn), Aluminium (Al), Cadmium (Cd), Lead (Pb), and Chromium (Cr), the samples were analyzed on Atomic Absorption Spec-trophotometer (Perkin Elmer) as shown in Figure 3.6 for concentration using specific cathode lamp. In AAS, the water sample was aspirated, aerosolized, and mixed with combustible gases (e.g. acetylene air, nitrous oxide), then vaporized and atomized in a flame at temperature of 2100 to 2800°C. The atoms in the sample were transformed into free, unexcited ground state atoms, which absorbed light at specific wavelengths. A light beam from a lamp whose cathode was made of the element of interest was passed through the flame. The amount of light absorbed was proportional to the concentration of the element in the sample [16].

## 8.3 RESULTS AND DISCUSSION

Table 1 showed the treated effluents and the untreated effluents.As shown, the treated effluents had good results compared to untreated effluent on pollutants. From Table 1, the untreated effluents and treated effluents had almost similar temperature which was 27.57°C and 27.83 °C respectively. Compared to the Environmental Quality (Industrial Effluents) Regulations 2009, the temperatures were considered low as the standard given was 40°C. Excessive temperature changes can accelerate chemical processes and can be detrimental to aquatic plants and wildlife. Increased heat in water can reduce its ability to hold dissolved oxygen, while sudden temperature 'shocks' (often caused by heated industrial water release into a lake or stream) can be harmful to many aquatic species [1].

From Table 1 the pH of the untreated effluents was 6.30 which was slightly acidic compared to the treated effluents which had the neutral pH of 7.44. Both pH values were within the permissible limits for industrial effluents set by Environmental Quality (Industrial Effluents) Regulations

**TABLE 1:** Physical Assessment of Semiconductor Waste water Effluents

| Parameters | Untreated Effluents | Treated Effluent |
|---|---|---|
| Temperature (°C) | 27.57 | 27.83 |
| pH | 6.30 | 7.44 |
| Turbidity (NTU) | 727.7 | 1.65 |
| Conductivity (mS/cm) | 0.241 | 0.44 |
| Total Suspended Solid (g) | 22.37 | 0.00 |

2009. High pH causes a bitter taste and low pH water will corrode or dissolve metals. A pH range of 6.0-9.0 is needed for healthy ecosystems. Sudden pH change often indicates chemical pollution [16].

Turbidity is a measure of the degree to which the water loses its transparency due to the presence of suspended particulates [17]. Turbidity in water is due to the occurrence of suspended matter which results in the scattering and absorption of light rays. Turbidity can be caused by phytoplankton or by sediments suspended in water. Water that is brown in colour is high in sediments while green, turbid water contains phytoplankton and other growths of aquatic life. It is well observed where the untreated effluents have the more suspended particulates that is 727.7 NTU whereas the treated effluents only with 1.65 NTU. High turbidity levels can indicate several problems for the water body. Turbidity blocks out sunlight needed by submerged aquatic vegetation. It also indicates low levels of dissolved oxygen caused by suspended solid [1].

The conductivity measured in the untreated effluents was 0.241mS/cm, which was lower than the treated effluents with value of 0.44mS/cm. High conductivity indicates the presence of high dissolved salt such as chloride, sulfate, sodium, calcium and others sources. Increases in this dissolved salt may affect the aquatic organisms [16].

The TSS values (Table 1) of the untreated effluents and treated effluents were 22.37g and 0.00g respectively. Both samples had low TSS values, but the treated effluents can be considered having no TTS value. Untreated effluents with that amount of TSS may cause handling problem, if this effluent is discharged to river or stream, it will make it unsuitable for aquatic life where the high amount of suspended solid will block the sunlight needed for the aquatic organisms to live and cause depletion of oxygen level.

The salinity measured for untreated effluents was 0.19 whereas for treated effluents was 0.07. High salinity (the presence of excess salts in water) can be harmful to certain plants, aquatic species, and human. High levels of salts in drinking water can lead to high blood pressure and other health concerns for humans.

While water molecules hold oxygen atom, this oxygen is not what is required by aquatic organisms living in our natural waters. A small quantity of oxygen, up to about ten molecules of oxygen per million of water, is actually dissolved in water. This dissolved oxygen is breathed by fish and zooplankton and is needed by them to live. Other gases can also be dissolved in water. In addition to oxygen, carbon dioxide, hydrogen sulfide and nitrogen are examples of gases that dissolve in water. Gases dissolved in water are significant. For example, carbon dioxide is important because of the role it plays in pH [18-19].

From Table 2, the untreated effluent had the value of 3.98 mg/L and the treated effluent had the value of 5.52mg/L. High levels of dissolved oxygen within the standard allow a variety of aquatic organisms to thrive [1] whereas many living organisms cannot survive in waters with DO levels of less than 1mg/L for more than a few hours.

BOD was determined by incubating a sealed sample of water for five days and measuring the loss of oxygen from the beginning to the end of the test. Samples often must be diluted prior to incubation or the bacteria will deplete all of the oxygen in the bottle before the test was completed. The main focus of wastewater treatment plants is to reduce theBOD in the effluent discharged to natural waters. Wastewater treatment plants are designed to function as bacteria farms, where bacteria are fed oxygen and organic waste. From the present investigation, the untreated effluents showedBOD of 31.51 mg/L whereas the treated effluents' was 23.42 mg/L. If effluent with high BOD levels is discharged into a stream or river, it will accelerate bacterial growth in the river and consume the oxygen levels in the river. The oxygen may diminish to levels that are lethal for most fish and many aquatic insects. As the river re-aerates due to atmospheric

**TABLE 2:** Chemical Assessment of Semiconductor Wastewater Effluents

| Parameters | Untreated Effluents | Treated Effluent |
|---|---|---|
| Salinity | 0.19 | 0.07 |
| Dissolved Oxygen (%) | 52.27 | 71.57 |
| Dissolved Oxygen (mg/L) | 3.98 | 5.52 |
| Biochemical Oxygen Demand (mg/L) | 31.51 | 23.42 |
| Chemical Oxygen Demand (mg/L) | - | 25.3 |
| Chlorine dioxide (mg/L) | 444.7 | 14.7 |

mixing and as algal photosynthesis adds oxygen to the water, the oxygen levels will slowly increase downstream [20]. The BOD of both treated and untreated effluents were below the standard level.

The COD test determined the oxygen required for chemical oxidation of organic matter with the help of strong chemical oxidant. The COD is a test which is used to measure pollution of domestic and industrial waste. The waste is measured in terms of equality of oxygen required for oxidation of organic matter to produce $CO_2$ and water. It is a fact that all organic compounds with a few exceptions can be oxidizing agents under the acidic condition. COD test is useful in pinpointing toxic condition and presence of biological resistant substances [21]. From Table 2, the COD value for untreated effluents was over the measuring range whereas for treated effluents was 25.3mg/L. Higher COD indicated higher amount of pollution in the wastewater and COD value was always greater than BOD values. The COD test is important to monitor and control the discharge of effluents and for assessing treatment plant performance as impact of effluents or wastewater discharge on river water is predicted by its oxygen demand.

Chlorine dioxide is an extremely effective disinfectant and bactericide, equal or superior to chlorine on a mass dosage basis. Its efficacy has been well documented in the laboratory, in pilot studies and in full-scale studies using potable and wastewater. Unlike chlorine, chlorine dioxide does not hydrolyze in water. Therefore, its germicidal activity is relatively constant over a broad pH range. Chlorine dioxide is as effective as chlorine in destroying coliform populations in wastewater effluents [22]. From Table 2, chlorine dioxide found in untreated effluent was 444.7 mg/L and in treated effluent was 14.7 mg/L. Low amount of chlorine dioxide is important as it acts as a disinfectant and might cause death to living organisms if in high amount.

Atomic Absorption Spectrophotometer (AAS) was used to quantify elements based on the amount of light that they absorb. For flame atomization, the resulting solution was nebulized to form fine droplets that were sprayed into the flame. A complex series of physical and chemical processes occurred to produce free gaseous atoms in the light path of the spectrometer. The amount of light absorbed was proportional to the concentration of the element in the solution. Flame AAS has detection limits at the parts-per-million level or mg/L. Every element absorbed and emit a unique set of wavelengths of light [23]. The heavy metals were elements with atomic weights between 63.5 and 200.5 and a specific gravity greater than 4.0 [17]. Chemical precipitation is most commonly employed for most of the metals. Typically, source reduction and stream segregation were practiced before these streams intermingle with others. Metals were precipitated as hydroxide through the addition of lime or caustic to a pH of minimum solubility.

Based from Table 3, the element of Cd and Cr were absent in both untreated and treated effluents. Cd is a relatively rare element where it was estimated to be present at an average concentration between 0.15 and 0.2 mg kg-1[24]. Cd is chemically very similar to Zn and they usually undergo geochemical processes together [25]. Al, Pb andMn were present together in the untreated effluents but after undergoing treatment, the water had zero value of these three elements. High concentration of Al in irrigation water can be toxic to plants whereas lead as a contaminant in water may come from commercial lead-containing products produced by the semiconductor company which will cause serious health problems to human and aquatic creatures as well. The amounts of Cu, Fe and Zn in the untreated effluents were high compared to the treated effluents. Cu and Fe were among the essential nutrients for plants, animals, and humans but high Cu concentrations are toxic whereas high Fe concentration may cause rusty color and metallic taste but it is not considered toxic. Zn, on the other hand will cause serious poisoning to humans if in high concentration [26].

## 8.4 CONCLUSION

The wastewater quality can be sustained within safe limits for better management of the plant. Industrial wastewater effluents are usually highly variable, with quantity and quality variations brought about by bath discharges, operation start-ups and shut-downs, working-hour distribution and so on. The design of the treatment facility is based on the study of the qualities that must be maintained in the environment to which the wastewater is to be discharged or for reuse of the wastewater and must follow the applicable environmental standards or discharge requirements.

**TABLE 3:** Concentration of Heavy Metal in Treated and Untreated Wastewater Effluents.

| Parameters | Untreated Effluents | Treated Effluent |
|---|---|---|
| Aluminium, Al (mg/L) | 2.33 | Nil |
| Copper, Cu(mg/L) | 1.65 | 0.01 |
| Cadmium, Cd(mg/L) | Nil | Nil |
| Chromium, Cr(mg/L) | Nil | Nil |
| Iron, Fe(mg/L) | >3.0 | 0.07 |
| Lead, Pb(mg/L) | 0.06 | Nil |
| Manganese, Mn(mg/L) | 0.10 | Nil |
| Zinc,Zn(mg/L) | >0.70 | 0.01 |

# REFERENCES

1.  Crittenden, J. C. (2005). Canada. John Wiley and Sons, INC.
2.  Crites, R. W., Reed, S. C., & Bastian, R. K. (2000). USA. McGraw-Hill.
3.  Chin, D. A. (2006). John Wiley and Sons, INC.
4.  Vibha Agrawal, S. A. Iqbal and Dinesh Agrawal., Orient. J. Chem., 26(4), 1345-1351 (2010)
5.  EPA, (1991). 1, 9-91.
6.  Rahman, A. A. A. (2009). Bachelor Science Thesis. Universiti Malaysia Pahang.
7.  Barnes, K. H., Meyer, J. L., & Freeman, B. J. (1998).Georgia Water Resources Conference, March 20-22, 1997, the University of Georgia, Athens Georgia
8.  Needleman, H., (2004). Lead poisoning, Annu. Rev. Med., 55: 209-212.
9.  Davis, J.M., D.A. Otto, D.E. Weil and L.D. Grant, (1999). Pub. Med – Indexed and MEDLINE, 12(3): 215-29.
10. HACH, (2010).Hach Company, Loveland, Colorado, USA.
11. HACH, (2009). Hach Company, USA
12. Jones, C., Bacon, L., Kieser, M.S., & Sheridan, D. (2006).USA. McGraw-Hill.
13. Cech, T. V. (2010).John Wiley and Sons, INC
14. Li, Y., & Migliaccio, K. (2011). Wastewater analysis. In Li, Y., & Migliaccio, K. (Eds.), Water Quality Concepts, Sampling, and Analysis, 1, 2-15. USA. CRC Press.
15. Kolhe, A. S., & Pawar, V. P. (2011). Recent Research in Science and Technology, 3, 29-32.

16. Li, Y., Zhou, M., & Zhao, J., (2011). Laboratory Qualifications for Water Quality Monitoring. In Li, Y., & Migliaccio, K. (Eds.), Water Quality Concepts, Sampling, and Analysis, 7, 137-156. USA. CRC Press.
17. Koranteng, A. E. J., Owusu, A. E., Boamponsem, L. K., Bentum, J. K., & Arthur, S. (2011). Pelagia Research Library, 2, 280-288.
18. Matani, A. G. (2006).Science Tech Entrepreneur. Retrieved from www.techno-preneur.net/information desk/sciencetech_magazine/
19. K. C. Gupta and J. Oberoi., Orient. J. Chem., 26(1), 215-221 (2010).
20. Manahan, S. E. (2010).USA. Taylor and Francis group, LLC.
21. Makhlough, A. (2008). Unpublished master's thesis.
22. Ahuja, S. (2009). Academic Press, 197-212.
23. Wang, L. K., Chen, J. P., Hung, Y. T., & Shammas, N. K. (2010). USA. Taylor and Francis group. LLC.
24. Manahan, S. E. (2010). USA. Taylor and Francis group, LLC.
25. Pennington, K.L. and Cech, T.V., (2010), Introduction of Water Resources and Environmental Issues, USA University Press.
26. Eckenfelder, W. W., Ford, D. L., & Englande, A. J. (2009).137-177. USA. McGraw Hill.

# CHAPTER 9

# Improving the Efficiency of a Coagulation-Flocculation Wastewater Treatment of the Semiconductor Industry through Zeta Potential Measurements

EDUARDO ALBERTO LÓPEZ-MALDONADO,
MERCEDES TERESITA OROPEZA-GUZMÁN, AND
ADRIÁN OCHOA-TERÁN

## 9.1 INTRODUCTION

In the industrial wastewater treatment of semiconductors, the most appreciated characteristic of polyelectrolytes (PE) is their solid-liquid separation efficiency, with extensive application in purification of drinking water, industrial raw and process water, municipal sewage treatment, mineral processing and metallurgy, oil drilling and recovery, paper and board production, and so forth [1–9]. Polyelectrolytes can be used alone or in association with other flocculant aids, such as inorganics salts, surfactants, or even as a second polymer. Compared with inorganic coagulants, there are some advantages by using organic polyelectrolytes [1, 6, 7, 10, 11]: lower dose requirements, a smaller volume of sludge, a smaller increase in the ionic load of the treated water, and cost savings up to 25–30%.

*Improving the Efficiency of a Coagulation-Flocculation Wastewater Treatment of the Semiconductor Industry through Zeta Potential Measurements.* © 2014 Eduardo Alberto López-Maldonado et al. Journal of Chemistry *Volume 2014 (2014), Article ID 969720 (http://dx.doi.org/10.1155/2014/969720).* Creative Commons Attribution License (http://creativecommons.org/licenses/by/3.0/).

It is well known that the efficiency of a certain polyelectrolytes in flocculation processes is evaluated as a function of four main parameters: the optimum flocculant concentration, which should be as low as possible; isoelectric point (IEP) that determines the effective pH range; and the flocculation window, which must be as large as possible [12]. To improve the solid/liquid separation process, polymeric flocculants, mainly polycations, have been used [1, 6–9]. The main disadvantage of flocculation with polymers is the very small flocculation window, risking particles resuspension with few dosage increases.

A number of studies have tried to solve this problem by combining two or more oppositely charged polyelectrolytes that can be added one after another [13–16] or as nonstoichiometric polyelectrolyte complexes (NIPECs) [17–24]. In the first case, a combination of a low molecular cationic weight and a high-molecular weight anionic polymer produce synergism during flocculation. This system is known as "dual flocculation process" [25]. For fine and ultrafine solid suspensions, use of double flocculant systems seems to offer a promising route for enhanced solid-liquid separation [25]. The dosing sequence, polymer size, and charge density all affect flocculation significantly. Enhanced flocs form through a combination of oppositely charged polyelectrolyte under suitable condition [13, 14, 22, 26, 27].

The investigations on NIPECs flocculants, as colloidal dispersions bearing positive or negative charges in excess, which started with the preliminary studies of Kashiki and Suzuki [17, 18] and were developed in the last years [19–24], have been concentrated on the use of NIPECs with molar ratio between charges ranged from 0.4 to 0.8. The main advantage in flocculation induced by NIPECs is the lower dependence on the concentration of the flocculants, showing a substantially wider optimum concentration range. Nevertheless, the optimum concentration required for flocculation with NIPECs was found to be higher than the optimum concentration for flocculation with polycations. Dual flocculation (sequential addition of two chemicals) presents some advantages compared with single polymer flocculant, as higher overall level of aggregation, less sensitivity to variations of polyelectrolyte concentration, good sludge dewatering, superior retention, and shear-resistant flocs, and so forth. All these issues improve control and optimization maneuvers of the flocculation process.

As mentioned above, in the semiconductor assembly industry the wastewater treatment system consists of a coagulation-flocculation process in which anionic and cationic polyelectrolytes are used to remove suspended solids, organic matter, and cation content so that the effluent meets the maximum permissible limits Pb 1 ppm, Ni 3.0 ppm, Cu 1.2 ppm, biochemical oxygen demand ($BOD_5$) 60 mg $O_2$/L, chemical oxygen demand (COD) 150 mg $O_2$/L, and total nitrogen (TN) 25 mg/L, prior to being discharged into the municipal sewer system and/or turned to be reused as service water in cooling towers, heat exchangers, and production.

One of the key stages in the production line of semiconductors is electroplating (EP). Its wastewater contains high concentration of suspended solids, organic compounds, and dissolved cation. Under these conditions, polyelectrolyte dosing should guarantee the entire maximum permissible concentration. Commonly operators in the wastewater treatment plant are used to modifying any independent variable as pH and PE dose to achieved the maximum allowable limit in total suspended solids (TSS) and cation content; however, the treated water is affected by another important parameter as COD, total organic carbon (TOC), and $BOD_5$.

In this work the authors decided to use real wastewater coming from the EP process, considering this water as the greater challenges for solid-liquid separation in the semiconductor industry. First of all, traditional coagulation-flocculation windows construction was followed with zeta potential measurements as well as turbidity, TSS, COD, and TOC to demonstrate that solid-liquid separation can be predicted by zeta potential measurements and could help to improve the coagulation-flocculation process efficiency, knowing the optimal polyelectrolyte dosage and opening a potential way for polyelectrolytes design ensuring low environmental impact of polyelectrolyte overdose.

## 9.2 EXPERIMENTAL

The experimental work was made in three steps. The first one was the characterization of electroplating raw wastewater under the Mexican environmental regulations NOM-002-SEMARNAT-1997 (Zn 1.2 ppm, Ni 3.0 ppm, Cu 1.2 ppm, Pb 1.0 ppm, TN 25 mg N/L, TSS 60 ppm, COD

150 mg $O_2$/L, and $BOD_5$ 60 mg O2/L), the second was to construct the pH-ζ diagrams allowed to establish different polyelectrolyte dosing strategies, and the third constructs coagulation-flocculation windows.

The experimental strategy followed to study the conditions of coagulation-flocculation process was to examine the profiles of pH and ζ of commercial polyelectrolytes used in assembly semiconductor wastewater treatment plant (WWTPs). Once the isoelectric point of the coagulant and flocculant was determined, as well as the sampled wastewater, polyelectrolytes dosing was studied at different pH values. In parallel the effect of using an interpolyelectrolyte complex flocculation in the window was tested. Finally, the effect of coagulant dose in the flocculation window by the dual flocculation process was performed. In Figure 1 the experimental methodology to evaluate the physicochemical performance of PE in coagulation-flocculation windows is resumed.

### 9.2.1 MATERIALS

Polydadmac (OPTIFLOC C-1008) and flocculant (Trident 27,506) are commercial polyelectrolytes used in a semiconductor industry.

All reagents used in testing water quality (COD, TN, TOC, and BOD5) were obtained from Hach as follows:

> Digesdahl Hach (microKendahl digestion apparatus),
> Hach Digital Reactor DRB200 (Digestor),
> Denver Instrument pH UP-5 (Potentiometer),
> DR/890 Hach (Colorimeter),
> HQ40d Multi Hach (measuring equipment for dissolved oxygen and conductivity),
> Dissolved Oxygen Meter (Luminescent Dissolved Oxygen) LD0101-01 Probe,
> Conductivity Meter CDC401-01 Probe,
> Atomic Absorption Spectrophotometer GBC 932 Plus,
> Zetasizer Nano-ZS, model ZEN3500.

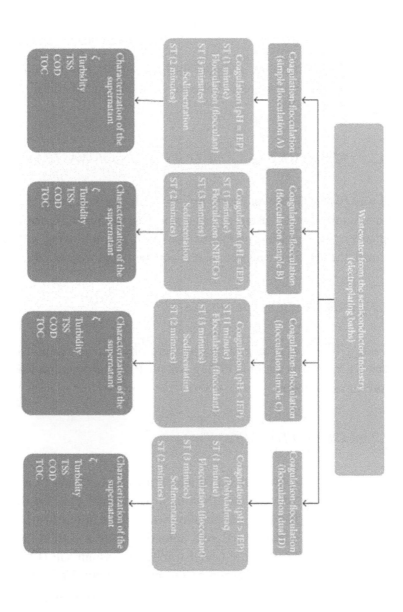

**FIGURE 1:** Experimental conditions for the physicochemical evaluation of polyelectrolytes used in coagulation-flocculation processes for electroplating wastewater in semiconductor industry.

## 9.2.2 METHODS

### 9.2.2.1 RESIDUAL WATER SAMPLING
### IN THE ASSEMBLING SEMICONDUCTOR INDUSTRY

The wastewater sampling protocol in the assembling semiconductor industry was followed as recommended by sampling Mexican standards (NMX-AA-003-1980).

### 9.2.2.2 PREPARATION OF THE DISPERSION OF WASTEWATER

The wastewater dispersions were prepared diluting 5 mL or raw wastewater in deionized water using a volumetric flask of 50 mL. Since the electroplating process is in continuous operation and following a timetable protocol, the sampled water content is considered as reproducible.

### 9.2.2.3 PREPARATION OF NIPECS

Anionic NIPECs synthetic solutions were prepared in relationships between 0.1 to 2 mg of cationic PE/mg anionic PE. Each solution was stirred before ζ measurement.

### 9.2.2.4 COLLOID TITRATION

Polydadmac 10 ppm was prepared taking a certain amount of a solution of 1.28 mM of polydadmac and diluting in 10 mL with distilled water in a volumetric flask. The prepared solution was poured into a 20 mL vial and measured to be entitled initial zeta potential. Then a certain amount of titrant solution 1.24 mM poly (vinyl sulfate) potassium salt (PVSK) was added. Each solution was stirred before measurement. Various additions of titrant were made to reach the isoelectric point (turbidity appears in solution).

## 9.2.2.5 MICRO-JAR TESTS

A sample of 5 mL of wastewater electroplating process was diluted with 50 mL of deionized water into a volumetric flask. pH was adjusted (5, 7, and 9). A second solution of wastewater with adjusted pH was prepared using a dilution factor of 0.5. The polyelectrolyte dosage tests (micro-Jar tests) were performed in 20 mL vials. Progressive additions of flocculant solution 0.1093 g/L were done and after each one, the vials were shaken for 2 min and allowed to settle for 2 more minutes. Finally, the supernatant was suctioned to determine turbidity, $\zeta$, TSS, TOC, and COD.

## 9.2.2.6 ANALYTICAL TECHNIQUES

Immediately, the samples were tested for $BOD_5$, COD, and TOC, five composite samples were analyzed following the procedures and test methods established by Mexican standards (NMX-AA-028-SCFI-2001, NMX-AA-030-SCFI-2001).

## 9.3 RESULTS AND DISCUSSIONS

## 9.3.1 CHARACTERIZATION OF WASTEWATER FROM ELECTROPLATING PROCESS IN A SEMICONDUCTOR INDUSTRY

The analysis of wastewater from different process units, in the assembling semiconductor industry, showed that electroplating section is the more challenging one. However, a detailed chemical composition is uncertain due to the large amount of chemical mixtures employed in the electroplating baths [28]. As shown in Table 1 electroplating wastewater has a high COD, turbidity, and total dissolved solids as well as Pb and Sn concentration.

Due to the fact that coagulation-flocculation processes are used in the installed WWTPs, the key parameters to follow-up are turbidity, TSS, and

**TABLE 1:** Characterization of electroplating wastewater.

| Parameter | ppm |
|---|---|
| Sn | 4854 |
| Pb | 1044 |
| Fe | 683 |
| TSS | 4510 |
| Turbidity (FAU) | 2990 |
| EC (mS/cm) | 74 |
| $\zeta$ (mV) | 45 |
| Size (nm) | 346 |
| Color | Milky |
| pH | 0.8 |
| COD (mg$O_2$/L) | 1432 |
| TOC (mgC/L) | 125 |
| BOD5 (mg$O_2$/L) | 30 |
| TN (mgN/L) | 50.6 |
| Biodegradability (BOD$_5$/COD) | 0.04 |

zeta potential ($\zeta$). All of them are directly related to the stability of the suspended solids. For raw wastewater $\zeta = 45$ mV (pH = 0.8), indicating the presence of positively charged particles suspended in water and probably related to metallic cations adsorbed on particles surface. For particles of 346 nm, longer sedimentation times are expected. Physical properties as turbidity and TSS also indicate that raw wastewater is a stable dispersion. Considering water properties, the solids separation using coagulation-flocculation processes seems to require the addition of negative polyelec-trolytes. However, the specific dose is unknown as is the real effect of negative polyelectrolyte addition on the solid separation. In this paper, traditional coagulation-flocculation process is analyzed to determine the optimal polyelectrolyte dose as well as the more suitable pH condition to improve the efficiency and to diminish the environmental impact caused by polyelectrolyte overdose.

## 9.3.2 DETERMINATION OF THE ISOELECTRIC POINT FOR THE ELECTROPLATING WASTEWATER

In the first stage (coagulation) suspended solids were destabilized by changing the water pH until isoelectric point is approached. A simple pH variation could be enough to stabilize or destabilize dispersions. Moreover, the performance of polymeric PE is influenced by the wastewater pH. Thus, the pH value may control both polyelectrolyte charge density and suspended particles surface charge. The isoelectric point of the dispersion generated in the electroplating process will be detected in a plot of $\zeta$ versus pH.

Using the colloidal titration method (with $\zeta$ as detection point) the charge density (CD) of the anionic and cationic PE (flocculant and coagulant) was determined: polydadmac 22 meq/g and flocculant 5 meq/g. These values confirm why polydadmac is usually a coagulant due to its high charge density, while the flocculant does not require high CD to accomplish the solid agglomeration. In Figure 2, the plot $\zeta$ versus pH shows the CD variation for suspended particles in wastewater, as well as the proper pH to achieve the isoelectric point. In the same plot, it can be observed that polydadmac does not reach an isoelectric point and flocculant has one at very low pH. This result implies that at pH > 5, flocculant is expected to be very efficient, while polydadmac is effective at pH > 10 (diminish of its CD is certainly due to the ammonium hydrolysis).

To corroborate the eye observable phenomena (TSS) with the interfacial phenomena (zeta potential) and nanoscale particle diameters, different combinations were studied and are presented below. Further studies were conducted to analyze the influence of PE dosage with organic matter content (COD and TOC).

The coagulation-flocculation experiments are generally divided into two parts: single flocculation (optimum condition for operation using only one PE) and dual PE flocculation.

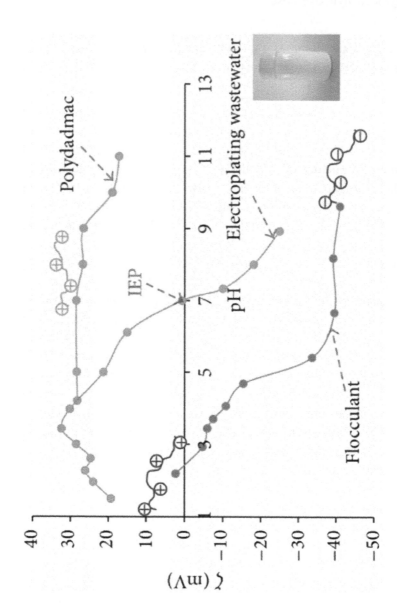

**FIGURE 2:** Electrokinetic properties of the dispersion of wastewater from electroplating, 20.8 ppm polydadmac, and 21.9 ppm flocculant.

## 9.3.3 SINGLE FLOCCULATION (A)

pH adjustment was used to reach the isoelectric point (pH 7); however, the settling kinetics was too slow and the flocculant addition was needed. Even if turbidity and TSS are the most common physical properties that guide the water clarification, they are not recommended for controlling dosage or investigate the source of an operation problem. To show this fact, in Figure 3 turbidity (FAU), TSS, $\zeta$, electric conductivity, TOC, and COD were plotted as a function of flocculant dosage in single flocculation of the neutralized wastewater. In these plots three main regions are identified: low dose, where at the flocculant concentration in the residual water does not allow the wastewater clarification; the optimal dose region, which corresponds to the optimum flocculant concentration for the total removal of TSS and turbidity; and the overdose region, indicating flocculant concentrations that cause stabilization of the dispersed particles and have an adverse effect on the quality of treated water. The magnitude Figure 6 of TSS and turbidity is the same (thousands) indicating that the generation of turbidity and TSS is due to the same cause.

Figure 3 indicates that at low dosages (2 and 13 mg/L) of flocculant, the $\zeta$ potential is approximately 8 mV and the turbidity is high (2587 FAU). At the point where the $\zeta$ potential value becomes −0.7 mV (15.6 ppm flocculant), the turbidity is at an absolute minimum. After this point in nearly obtaining a zero charge and turbidity, further addition of flocculating agent now reverses and increases the charge of the contaminating material ($\zeta$ = −6.3 mV and turbidity = 2630 FAU) and therefore restabilizes these particles in the water. In this region of overdose, the polyelectrolyte in excess causes the adsorption of polyelectrolyte chains onto the stable particles in suspension, having an adverse effect on water quality. This is observed by increasing the turbidity and suspended solids as zeta potential become more negative. The shift of the value of $\zeta$ is due to the negative charge of the flocculant. Electric conductivity of water remains virtually constant after the first additions of flocculant. For the range between 2 and 13 mg/L, the supernatant has small but positive zeta potential (8 mV). Figure 3 also shows the variation in the TOC and COD content in the supernatant. It is noted that since the first dose of flocculant (2 ppm) COD at 1500 decreases

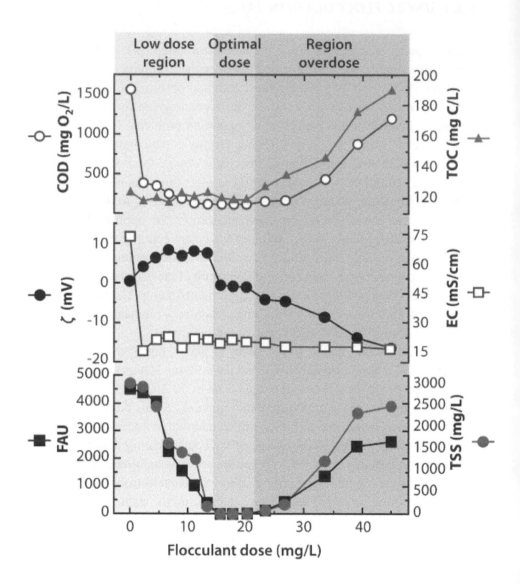

**FIGURE 3:** Turbidity, TSS, ζ, EC, COD, and TOC of the supernatant in the single flocculation at pH 7 with flocculant.

to a value of 380 ppm, subsequently, a gradual COD decrease occurs with increasing flocculant dose until the optimal dose of 15.6 mg/L.

It is also noted that the TOC is almost constant in the interval of the flocculation window. At the point of optimum flocculant dose (15.6 mg/L) the supernatant has a COD equal to 122 mg /L and TOC equal to 119 mg C/L, which indicates that the treatment applied do not remove residual organic matter. Figure 4 shows the rate of sedimentation of suspended solids in each dose of flocculant. The relationship between flocculant dose and sedimentation speed presents a maximum. In the region of overdose, the sedimentation rate decreases by restabilization of the suspended solids.

## 9.3.4 NIPECS SIMPLE FLOCCULATION (B)

Various materials have been developed in recent years for coagulation and flocculation purposes, as inorganic-based coagulants, organic-based flocculants, and hybrid materials [29]. The continuous increase of market needs for more efficient and effective materials in wastewater treatment has induced the development of hybrid materials for coagulation-flocculation of wastewater. Hybrid materials thus have emerged as new materials that pose tremendous potential in treating wastewater due to their better performance compared to that of conventional inorganic-based coagulants and its lower cost than that of organic-based flocculants [30].

Due to the synergetic effect of hybrid components in one material, hybrid materials hence pose a superior performance than that of individual component [31–33]. Compared with individual coagulant/flocculant, hybrid materials, which have combined functional components into one prescription, would be a convenient alternative material for the operation of wastewater treatment facilities since the whole wastewater treatment can be conducted with the addition of one chemical and in one tank instead of two unit operations in the conventional coagulation-flocculation system. Reduction of operation time as a result of the application of these hybrid materials in a single operation is favorable to the industries that are discharging large volumes of wastewater [34].

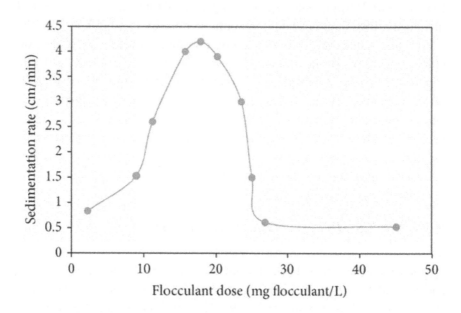

**FIGURE 4:** Sedimentation rate of suspended solids.

Within the hybrid materials currently used to improve coagulation-flocculation processes are the nonstoichiometric polyelectrolyte complexes (NIPECs).

Figure 5 shows the variation of $\zeta$ and size of the colloidal particles formed according to the relationship mg polydadmac/mg flocculant. It can be seen that the zeta potential becomes more positive as the amount of cationic polyelectrolyte increases, while the size remains constant as far as mg polydadmac/mg flocculant varies from 0.9 to 1.5. At ratios greater than 1.6, the colloidal particle size increases suddenly. The colloidal particles obtained in a ratio of 1.7 are neutral and have a size of 1.5 microns, and at this point the isoelectric point is reached. Interpolyelectrolyte complexes formed in ratios greater than 1.8 have a size of 2 mm and an excess of positive charge is reflected by the zeta potential value of 15 mV. During all these experiments, the dispersions characteristics remained constant indicating that changes of $\zeta$ and particle size variations were due to the NIPECs concentration and not for agglomeration kinetics.

## 9.3.4.1 PROPOSED MECHANISM FOR THE FORMATION OF NIPECS

Taking into account the DLS and $\zeta$ potential results, a mechanism in four steps has been assumed and presented in Figure 6.

In the first step, the added polydadmac interacts with flocculant chains leading to the primary aggregates, which, taking into account the differences in the flexibility of the complementary polyions and the mismatch of charges, may contain more flocculant chains connected by fewer polydadmac chains; such aggregates would have a high density of free negative charges compensated with small counterions, not by polydadmac charges.

The further addition of polydadmac (step II, Figure 6) led to the step-by-step neutralization of the negative charges of flocculant included in the primary aggregates, accompanied by rearrangements of chains and the formation of more compact particles with lower sizes. This assumption is supported by the monotonous decrease of the particle sizes found by DLS measurements with an increase of the mass ratio mg polydadmac/mg flocculant up to about 0.9.

For the polyion pair investigated in this work, it was observed by both DLS and $\zeta$ potential (Figure 6), that for a dose ranging from 0.9 up to 1.5 mg polydadmac/mg flocculant, the particle size remained almost constant (step III, Figure 6). It seems that the ratio of charge of about 1.7 is critical for these systems, because an abrupt increase of the particle sizes (secondary aggregation) and decrease in $\zeta$ were observed after this ratio (step IV, Figure 6).

Taking acount the results discussed above, NIPECs were prepared at a ratio of 1.5 mg polydadmac/mg flocculant with the following characteristics: particle size 132 nm and $\zeta = -25$ mV. The formed complexes were used as a new flocculant in destabilizing the dispersed solids of the wastewater from electroplating process. Figure 7 shows the variation of

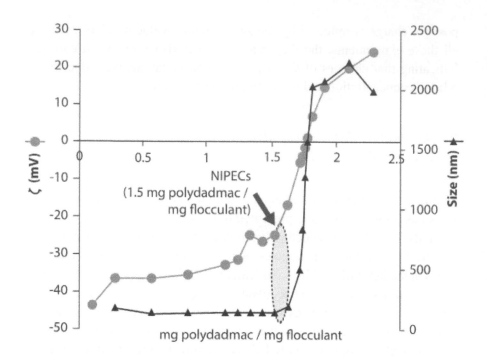

**FIGURE 5:** NIPECs of polydadmac/flocculant.

residual TSS and $\zeta$ in the supernatant according to NIPECs dose (mg/L). The dispersion of NIPECs is more effective in removing suspended solids and has a wider flocculation window (24 mg/L to 35.6 mg/L of NIPECs) compared with the pure flocculant. NIPECs dose required (greater than 41 mg/L) for the restabilization of dispersed particles, in which the value is $\zeta = -8.6$ mV, is higher in comparison with the dose of flocculant in system (A) (see above).

At a dose of 6 mg/L of NIPECs, COD is reduced from 1558 to 360 mg/L and decreases more in a range between 18 and 35 mg/L COD, while TOC is kept constant. The lower COD value of 132 mg/mL is achieved with a dose of 35 mg/L of NIPECs. At doses greater than 41 mg/L of NIPECs the restabilization of dispersed solids is induced. Again TOC increased corresponding to the increase in the concentration of residual NIPECs (see Figure 7).

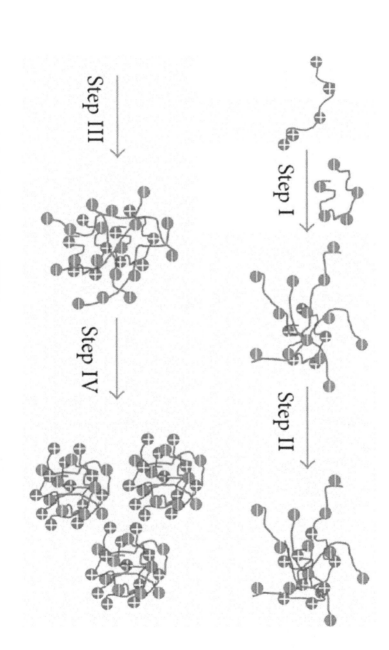

**FIGURE 6:** Proposed mechanism for the formation of NIPECs.

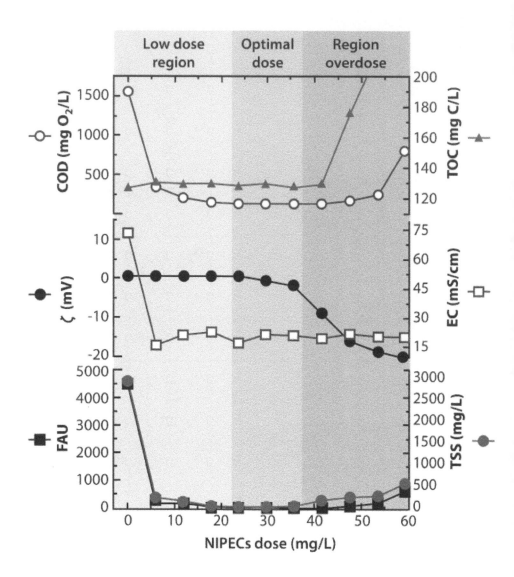

**FIGURE 7:** Turbidity, TSS, ζ, EC, COD, and TOC of the supernatant in the single flocculation at pH 7 with NIPECs.

## 9.3.5 SINGLE FLOCCULATION AT pH < IEP (C)

The residual water at pH 5 has positive $\zeta$ = 11 mV. At this pH flocculant has higher charge density (Figure 2). Figure 8 shows the variation of TSS content in the supernatant as a function of the flocculant dose. It is observed that in the region of overdose, the increase of dispersed solids has a smaller slope compared with the restabilization region of the single flocculation at pH 7 (A). The optimal dose corresponds to 33.5 mg/L of flocculant, greater than case (A). The flocculation window was between 33.5 and 40 ppm of flocculant.

## 9.3.6 DUAL FLOCCULATION (D)

The residual water at pH 9 has $\zeta$ = −28 mV. Therefore, the addition of flocculant at this pH has no meaning because the flocculant provokes repulsive interaction that restabilizes particles. Since at pH 9 the maximum cation removal is achieved, it was decided to perform the process of coagulation using the polydadmac.

In this case, the $\zeta$ variation versus polydadmac dose confirms the proper performance of the dual flocculation process, wherein the optimum condition of the coagulation step is achieved with a dose of 162 mg/L. Complementary, Figure 9 shows the EC and $\zeta$ of the resulting supernatant for each dose of polydadmac. It is noted that the dosage of polydadmac decreases the surface charge of the dispersed particles and the isoelectric point is reached at 162 mg/L of polydadmac. At doses greater than 162 ppm, inversion of surface charge of the dispersed particles was observed. The addition of polydadmac in the coagulation step causes a decrease in the EC; this is attributed to the removal of the anions present in the water that interact with positive sites in the quaternary polydadmac chains (Figure 9).

The next step after finding the optimal dose of polydadmac is the addition of flocculant. Figure 10 shows that a dose of 3.8 mg/L of flocculant reduced COD from 1432 mg/L to 380 mg/L, whereas the TOC remains

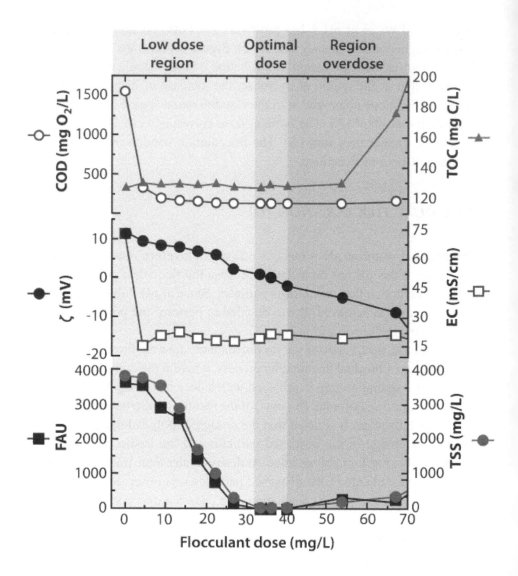

**FIGURE 8:** Turbidity, TSS, $\zeta$, EC, COD, and TOC of the supernatant in the single flocculation at pH 5 with flocculant.

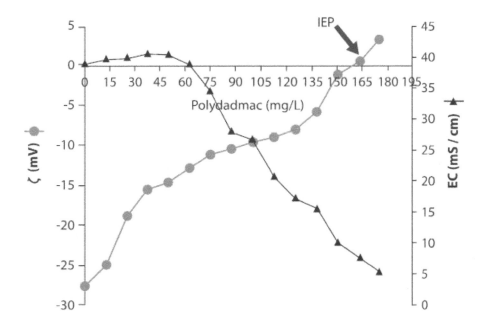

**FIGURE 9:** Effect of dose of polydadmac in ζ and EC the supernatant of the wastewater.

constant. Increasing flocculant doses progressively reduces the COD. At 18.8 mg/L TOC decreased to 119 mg/L.

At a dose of 26.4 mg/L of flocculant, which corresponds to the optimum dose for the removal of TSS, TOC decreased to 115 mg/L. In the range of 26.4 mg/L to 67 mg/L of flocculant, TOC decreased markedly; it is noteworthy that in this dosage range metal content decreased to comply with regulations. At the dose of 67 mg/L the maximum TOC and COD attaint 45 mg/L. The dose of 77 mg/L of flocculant has an adverse effect on water quality, observed in increasing TSS and cations; this overdose of flocculant causes the increased value of ζ, COD, and TOC. The increase of

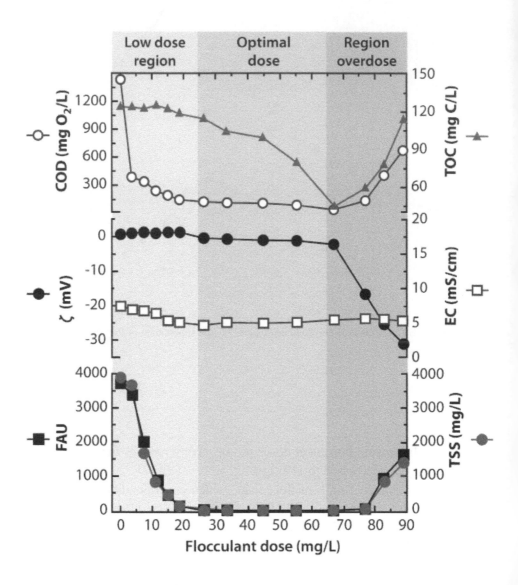

**FIGURE 10:** Turbidity, TSS, ζ, EC, residual COD, and TOC of the supernatant in the dual flocculation at pH 9 using polydadmac and flocculant.

COD is mainly attributed to the redispersion of solids corresponding to the inorganic material, while increasing TOC is attributed to increased residual flocculant concentration in solution (organic matter). The coagulation-flocculation process under described conditions removed OM up to 37%.

## 9.3.7 COMPARISON OF FLOCCULATION WINDOWS

A resume of all four strategies tested with wastewater from electroplating section in an assembly semiconductor industry is reported in Table 2. Results indicate that flocculation windows are determined not only by pH value but also by the combination of commercial PE doses and addition sequence.

Finally, Figure 11 presents the comparison of all flocculation windows. This representation is useful to select the ranges of flocculant doses under different pH conditions and additions strategies. Analysis of $\zeta$ potential versus coagulant dosage results is used to evaluate the effectiveness of various polyelectrolytes. Knowledge of $\zeta$ potential is important to adjustment of coagulant dosage levels periodically in order to minimize the cost of chemicals for wastewater treatment.

It was observed that at pH 5 the optimum dose is 33.5 mg/L and the flocculation dosage window is very short 33.5 to 40.0 mg/L of flocculant.

**TABLE 1:** Characterization of electroplating wastewater.

| Treatment system | pH | Optimal dose (ppm) | Flocculation window (ppm) |
|---|---|---|---|
| (A) Single flocculation | 5 | 33.5 | 33.5 to 40.2 |
| (B) Single flocculation | 7 | 15.6 | 15.6 to 20.1 |
| (C) NIPECs flocculation | 7 | 23.7 | 23.7 to 35.6 |
| (D) DUAL coagulation Flocculation | 9 | 162 | 26.4 to 67 |
| | | 26.4 | |

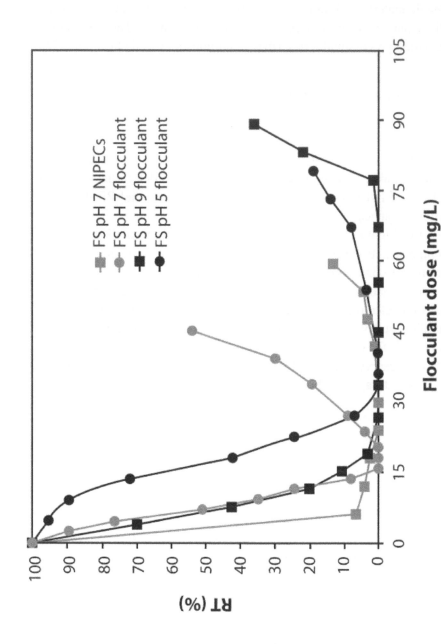

**FIGURE 11:** Flocculation windows comparison between different systems treatment.

The single flocculation window at pH 7 using the flocculant is much shorter (15.6 to 20.0 mg/L) than at pH 5; the redispersion zone is steeper than flocculation at pH 5. The optimal dose of flocculant at a pH of 7 is lower by 50% than that required for the simple flocculation at pH of 5. Flocculation window obtained in simple flocculation process at pH 7 using NIPECs is flattened compared to the previous two, and since the first dosage of NIPECs considerably a decreased in turbidity is achieved.

Flocculation window corresponds to NIPECs that is 24.0 mg/L to 35.6 mg/L, which is much wider than a pH of 5 and 7 using only the flocculant.

The flocculation window in wastewater treatment by dual flocculation at pH 9 is displaced to higher flocculant dose (26.4 mg/L to 67.0 mg/L of flocculant). Unlike using NIPECs flocculation window, over 8.1 ppm interpolyelectrolyte complex is required to achieve approximately the same %RT compared to the optimum dose of the flocculant used in the simple flocculation at pH 7.

## 9.4 CONCLUSIONS

The construction of a pH-$\zeta$ curve for wastewater, cationic PE, and anionic PE is the first step previous to selecting PE type and dose in a coagulation-flocculation process as those used in the semiconductor industry. This type of plot allows identifying the isoelectric point for each polyelectrolyte and wastewater. The flocculant has its isoelectric point at a pH of 2.5, at which pH > IEP interacts with the dispersed particles with positive surface charge.

Measurement of COD, TOC, TSS, and turbidity during the coagulation-flocculation windows construction allowed determining the separation efficiencies and the environmental impact. Zeta potential measurements served as a physicochemical evidence of the coagulation-flocculation efficiency in micro-Jars experiments.

At the pH value of 7, the flocculation window is 15.6 to 20.1 ppm and any higher flocculant dose at 15.6 ppm, due to the fact that the high charge density of polyelectrolyte has a negative effect on the efficiency of removal of metal hydroxides formed by the restabilization of suspended solids.

The use of the flocculant to pH 5 in the removal of suspended particles is effective but requires a much higher dose (33.5 ppm flocculant) compared to a required pH of 7 (15.6 ppm flocculant), due to the decrease of the charge density of the flocculant with respect to pH.

The main advantage of NIPECs is the substantial extension of the flocculation window. However, the optimal NIPECs concentration required for flocculation is higher (23–35 ppm NIPECs) than that required when using the flocculant only. The complexes formed in a ratio of 1.5 mg polydadmac/mg flocculant were tested in the destabilization of the dispersion of residual water. The NIPECs flocculants are more effective than single anionic polyelectrolyte; the critical concentration for restabilizing suspended solids is much larger for NIPECs (35 ppm).

The wastewater treatment by dual flocculation at pH 9 (162 ppm corresponding to PIE polydadmac residual water and 67 ppm of flocculant) manages content efficiently to remove turbidity, organic matter, and suspended solids up to 37%.

Every single flocculation system addition (flocculant or NIPECs) had the effect of diminishing COD, achieving 125 mg $O_2$/L for the optimal dose of each system. Complete removal of TSS and turbidity is achieved at the optimal dose of polyelectrolyte, while TOC remained roughly constant at 120 mg C/L.

## REFERENCES

1. B. Bolto and J. Gregory, "Organic polyelectrolytes in water treatment," Water Research, vol. 41, no. 11, pp. 2301–2324, 2007.
2. C. Negro, E. Fuente, L. M. Sánchez, Á. Blanco, and J. Tijero, "Evaluation of an alternative flocculation system for manufacture of fiber-cement composites," Industrial & Engineering Chemistry Research, vol. 45, no. 20, pp. 6672–6678, 2006.
3. A. H. Mahvi and M. Razavi, "Application of polyelectrolyte in turbidity removal from surface water," American Journal of Applied Sciences, vol. 2, no. 1, pp. 397–399, 2005.

4.  T. Tripathy and B. Rajan De, "Flocculation: a new way to treat waste water," Journal of Physical Sciences, vol. 10, pp. 93–127, 2006.
5.  S. Dragan, A. Maftuleac, I. Dranca, L. Ghimici, and T. Lupascu, "Flocculation of montmorillonite by some hydrophobically modified polycations containing quaternary ammonium salt groups in the backbone," Journal of Applied Polymer Science, vol. 84, no. 4, pp. 871–876, 2002.
6.  L. Ghimici, I. A. Dinu, and E. S. Dragan, "Application of polyelectrolytes in phase separation processes," in New Trends in Ionic (Co) Polymers and Hybrids, pp. 31–64, Nova Science, Hauppauge, NY, USA, 2007.
7.  D. Zeng, J. Wu, and J. F. Kennedy, "Application of a chitosan flocculant to water treatment," Carbohydrate Polymers, vol. 71, no. 1, pp. 135–139, 2008.
8.  F. Renault, B. Sancey, P.-M. Badot, and G. Crini, "Chitosan for coagulation/flocculation processes—an eco-friendly approach," European Polymer Journal, vol. 45, no. 5, pp. 1337–1348, 2009.
9.  S. Bratskaya, V. Avramenko, S. Schwarz, and I. Philippova, "Enhanced flocculation of oil-in-water emulsions by hydrophobically modified chitosan derivatives," Colloids and Surfaces A: Physicochemical and Engineering Aspects, vol. 275, no. 1–3, pp. 168–176, 2006.
10. J. R. Rose and M. R. St. John, "Flocculation," in Encyclopedia of Polymer Science and Engineering, J. I. Kroschwitz, Ed., vol. 7, pp. 211–233, John Wiley & Sons, New York, NY, USA, 2nd edition, 1987.
11. J. Gregory, "The stability of solid-liquid dispersions in the presence of polymers," in Solid/liquid Dispersions, T. F. Tadros, Ed., pp. 163–181, Academic Press, London, UK, 1987.
12. C. Laue and D. Hunkeler, "Chitosan-graft-acrylamide polyelectrolytes: synthesis, flocculation, and modeling," Journal of Applied Polymer Science, vol. 102, no. 1, pp. 885–896, 2006.
13. X. Yu and P. Somasundaran, "Enhanced flocculation with double flocculants," Colloids and Surfaces A: Physicochemical and Engineering Aspects, vol. 81, pp. 17–23, 1993.
14. A. Fan, N. J. Turro, and P. Somasundaran, "A study of dual polymer flocculation," Colloids and Surfaces A: Physicochemical and Engineering Aspects, vol. 162, no. 1–3, pp. 141–148, 2000.
15. G. Petzold, M. Mende, K. Lunkwitz, S. Schwarz, and H.-M. Buchhammer, "Higher efficiency in the flocculation of clay suspensions by using combinations of oppositely charged polyelectrolytes," Colloids and Surfaces A: Physicochemical and Engineering Aspects, vol. 218, no. 1–3, pp. 47–57, 2003.
16. B. U. Cho, G. Garnier, T. G. M. van de Ven, and M. Perrier, "A bridging model for the effects of a dual component flocculation system on the strength of fiber contacts in flocs of pulp fibers: implications for control of paper uniformity," Colloids and Surfaces A: Physicochemical and Engineering Aspects, vol. 287, no. 1–3, pp. 117–125, 2006.

17. I. Kashiki and A. Suzuki, "On a new type of flocculant," Industrial & Engineering Chemistry Fundamentals, vol. 25, no. 1, pp. 120–125, 1986.

18. A. Suzuki and I. Kashiki, "Flocculation of suspension by binary (polycation-polyanion) flocculant," Industrial & Engineering Chemistry Research, vol. 26, no. 7, pp. 1464–1468, 1987.

19. H. M. Buchhammer, G. Petzold, and K. Lunkwitz, "Salt effect on formation and properties of interpolyelectrolyte complexes and their interactions with silica particles," Langmuir, vol. 15, no. 12, pp. 4306–4310, 1999.

20. H.-M. Buchhammer, G. Petzold, and K. Lunkwitz, "Nanoparticles based on polyelectrolyte complexes: effect of structural and net charge on the sorption capability for solved organic molecules," Colloid and Polymer Science, vol. 278, no. 9, pp. 841–847, 2000.

21. R. S. Nyström, J. B. Rosenholm, and K. Nurmi, "Flocculation of semidilute calcite dispersions induced by anionic sodium polyacrylate-cationic starch complexes," Langmuir, vol. 19, no. 9, pp. 3981–3986, 2003.

22. G. Petzold, U. Geissler, N. Smolka, and S. Schwarz, "Influence of humic acid on the flocculation of clay," Colloid and Polymer Science, vol. 282, no. 7, pp. 670–676, 2004.

23. S. Schwarz and E. S. Dragan, "Nonstoichiometric interpolyelectrolyte complexes as colloidal dispersions based on NaPAMPS and their interaction with colloidal silica particles," Macromolecular Symposia, vol. 210, no. 1, pp. 185–191, 2004.

24. M. Mende, S. Schwarz, G. Petzold, and W. Jaeger, "Destabilization of model silica dispersions by polyelectrolyte complex particles with different charge excess, hydrophobicity, and particle size," Journal of Applied Polymer Science, vol. 103, no. 6, pp. 3776–3784, 2007.

25. G. Petzold, S. Schwarz, and K. Lunkwitz, "Higher efficiency in particle flocculation by using combinations of oppositely charged polyelectrolytes," Chemical Engineering & Technology, vol. 26, no. 1, pp. 48–53, 2003.

26. X. Yu and P. Somasundaran, "Role of polymer conformation in interparticle-bridging dominated flocculation," Journal of Colloid and Interface Science, vol. 177, no. 2, pp. 283–287, 1996.

27. H. Ono and Y. Deng, "Flocculation and retention of precipitated calcium carbonate by cationic polymeric microparticle flocculants," Journal of Colloid and Interface Science, vol. 188, no. 1, pp. 183–192, 1997.

28. S. H. Lin and C. D. Kiang, "Combined physical, chemical and biological treatments of wastewater containing organics from a semiconductor plant," Journal of Hazardous Materials, vol. 97, no. 1–3, pp. 159–171, 2003.

29. P. A. Moussas and A. I. Zouboulis, "A new inorganic-organic composite coagulant, consisting of Polyferric Sulphate (PFS) and Polyacrylamide (PAA)," Water Research, vol. 43, no. 14, pp. 3511–3524, 2009.

30. Y. Wang, B. Y. Gao, Q. Y. Yue, J. C. Wei, and W. Z. Zhou, "Novel composite flocculent polyferric chloride-polydimethyldiallylammonium chloride (PFC-PDMADAAC): its characterization and flocculation efficiency," Water Practice and Technology, vol. 1, no. 3, pp. 1–9, 2006.

31. N. D. Tzoupanos and A. I. Zouboulis, "Preparation, characterisation and application of novel composite coagulants for surface water treatment," Water Research, vol. 45, no. 12, pp. 3614–3626, 2011.

32. H. Tang and B. Shi, "The characteristics of composite flocculants synthesized with polyaluminium and organic polymers, chemical water and wastewater treatment VII," in Proceedings of the 10th Gothenburg Symposium, H. H. Hahn and E. Hoffmann, Eds., pp. 17–28, Gothenburg, Sweden, 2002.

33. W. Y. Yang, J. W. Qian, and Z. Q. Shen, "A novel flocculant of Al(OH)3-polyacrylamide ionic hybrid," Journal of Colloid and Interface Science, vol. 273, no. 2, pp. 400–405, 2004.

34. K. E. Lee, T. T. Teng, N. Morad, B. T. Poh, and Y. F. Hong, "Flocculation of kaolin in water using novel calcium chloride-polyacrylamide (CaCl2-PAM) hybrid polymer," Separation and Purification Technology, vol. 75, no. 3, pp. 346–351, 2010.

[3] R. Y. L.....

# PART IV

# PULP AND PAPER INDUSTRIES

Pulp and paper mills are considered one of the most polluting industries in the world. Wastewater discharged from pulp and paper mills contains solids, dissolved organic matter such as lignin, and inorganic materials such as chlorates and metal compounds. Nutrients such as nitrogen and phosphorus also in the wastewater can cause or exacerbate eutrophication in freshwater bodies, and other pollutants can be deadly to human, animal, and plant life.

CHAPTER 10

# Improvement of Biodegradability Index through Electrocoagulation and Advanced Oxidation Process

A. ASHA, KEERTHI SRINIVAS, A. MUTHUKRISHNARAJ, AND N. BALASUBRAMANIAN

## 10.1 INTRODUCTION

Pulp and paper industry is a highly capital, energy, and water intensive industry, also a highly polluting process and requires substantial investments in pollution control equipments. Looking into the serious nature of pollution, the pulp and paper industry in India has been brought under the 17 categories of highly polluting industries. India produces 6 million tonnes of paper per year though 311 mills by consuming around 900 million m3 of water and discharging 700 million $m^3$ of wastewater. Out of these about 270 small paper mills (capacity $\leq$10,000 tonnes per annum (TPA), having a total installed capacity of 1.47 MTPA) do not have chemical recovery units [1]. Effluents from this industry cause alternations in hydrographical parameters of the water body thereby causing tremendous

*Improvement of Biodegradability Index Through Electrocoagulation and Advanced Oxidation Process.* © 2014 A. Asha, Keerthi, A. Muthukrishnaraj, and N. Balasubramanian; licensee SpringerLink. com. *International Journal of Industrial Chemistry March 2014, 5:4, 10.1007/s40090-014-0004-x.*

to the ecosystem. The sources of pollution among various process stages in pulp and paper industry are wood preparation, pulping, pulp washing, bleaching, and paper machine and coating operations. Common pollutants include suspended solids, oxygen demanding wastes, colour, basicity, heavy metals, alkali and alkaline earth metals, phenols, chloro-organics, cyanide, sulphides and other soluble substances [2]. Recent progress in the treatment of persistent organic pollutants in wastewater has led to the development of advanced oxidation processes (AOPs). Advanced oxidation processes make use of strong oxidants to reduce COD/BOD levels, and to remove both organic and oxidizable inorganic components. The processes can completely oxidize organic materials to carbon dioxide and water with the help of free hydroxyl radicals ($OH\cdot + OH-$). Advanced oxidation process offer several advantages like process operability, absence of secondary waste and soil remediation. This method of treatment can be used either as a main treatment or as a hybrid technique [3]. It can also be used as a pre-treatment scheme for difficult wastewater for which feasible treatment methods are not available.

There are various methods available for treatment which includes biological method; physical methods like adsorption, membrane filtration; and chemical oxidation methods like electro-oxidation [4]. In advanced oxidation technique formation of strong oxidants plays an important role for the breakdown of pollutants into simple compounds [5]. Hydroxyl radicals can be produced by various methods such as electro-oxidation, photochemical and ozonation. Effectiveness of techniques is proportional to the ability to generate hydroxyl radicals.

In this present investigation, three different methods, i.e. electrocoagulation, electro-oxidation and photochemical methods were carried out to treat pulp and paper effluent. Electrocoagulation process is the in situ production of coagulants by means of electrolysis. Electro-oxidation is a process of mineralizing pollutants by electrolysis using anodes. It is of two types: direct oxidation and indirect oxidation. In direct method, anodic electron transfer takes place similar to chemical oxidation on anodic surface. In indirect method demineralization takes place in the presence of ferric and chloride ions [6]. Photolysis involves the interaction of light with molecules to bring about their dissociation into fragments. The addition of energy as radiation to a chemical compound is the principle of photochemical processes. Molecules absorb this energy and reach excited states with decay times long enough to take part in chemical reactions.

In this present investigation, improvement of biodegradability index using electrocoagulation and different advanced oxidation processes was studied. The influence of individual parameters on treatment was analysed. Significance of the method was also analysed and reported.

## 10.2 MATERIALS AND METHODS

The pulp and paper wastewater was used for the present investigation. The characteristics of the wastewater are given in the Table 1. Reactor consists of electrodes of 4.5 × 5.5 cm in size where mild steel was used as anode for electrocoagulation and graphite for electro-oxidation. Stainless steel was used as cathode for both the processes. Experiments were carried out in a batch electrochemical reactor of 250 ml capacity. The active surface area of the electrode was 25 cm$^2$ and anode–cathode distance was maintained at 1 cm. 1 g l$^{-1}$ of sodium chloride was used as

TABLE 1: Chemical Assessment of Semiconductor Wastewater Effluents

| Parameter | Value |
|-----------|-------|
| COD | 770 mg l$^{-1}$ |
| BOD | 105 mg l$^{-1}$ |
| BI | 0.13 |
| TDS | 2,160 mg l$^{-1}$ |
| TSS | 102 mg l$^{-1}$ |
| Chloride as Cl | 31 / mg l$^{-1}$ |
| Sulphate as SO$_4$ | 368 mg l$^{-1}$ |

supporting electrolyte. For mixing the reactor contents, a 1.5-cm-long stirring bead was used and the reactor was placed over a magnetic stirrer. The DC power supply system used was capable of supplying DC power in the range of 0–32 V/0–10 A.

Photochemical reactor setup consists of UV irradiation source as an 8-Watt lamp and its maximum emission is at 365 nm. Experiments were carried out in a batch photochemical reactor. Effluent has been taken in sample tube of capacity 0.1 L. $H_2O_2$ was used as an oxidant, which was added externally to enhance the efficiency of the photochemical process. The samples were irradiated for a period of 3 h (180 min) with a sampling interval of 30 min. The sample was immediately analysed for percentage COD removal.

## 10.3 RESULTS AND DISCUSSION

The wastewater from pulp and paper industry was collected and treated using three different techniques namely electrocoagulation, electro-oxidation, photochemical methods. Each treatment techniques were optimized with respect to various operating conditions. Comparisons between techniques have been done depending on improvement of biodegradability index with process time.

### 10.3.1 ELECTROCOAGULATION

Electrocoagulation is a synergistic process with a complex mechanism operating to remove pollutants from the water. Electrocoagulation operates by the dissolution of metal from the anode and with simultaneous formation of hydroxyl ions and evolution of hydrogen gas at the cathode. The hydroxide flocculates and coagulates the suspended solids thereby purifying the water. The generated ferric ions form monomeric, ferric hydroxo complexes with hydroxide ions and polymeric species, depending on the pH range. The $Fe(OH)_3$ flocs capture the pollutant molecule present in the wastewater to form sludge as shown in the following reaction:

$$Pollutant + Fe(OH)_{3(s)} \rightarrow [Sludge] \qquad (1)$$

To optimize the electrocoagulation operating parameters, the reduction of COD with electrolysis time has been analysed. Experiments were carried out at different pH, current densities and supporting electrolyte. Figure 1 shows the influence of pH on %COD removal. Reduction in COD percentage was high at neutral pH. At higher value of pH, COD reduction was not significant. This is due to the reason that at higher pH value, the solubility of complex formed increases which do not involve in the COD removal [7].

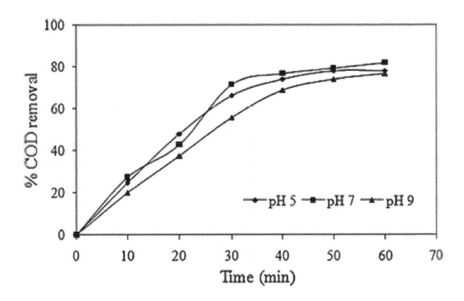

**FIGURE 1:** Effect of pH on percentage COD removal with electrolysis time; anode: mild steel; current density 10 mA cm$^{-2}$

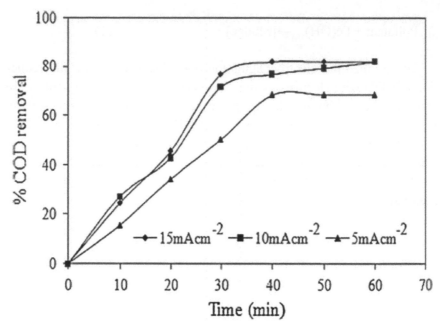

**FIGURE 2:** Variation of percentage COD removal with current density; anode: mild steel; pH: 7

Figure 2 depicts the influence of various current densities on COD reduction. It has been noted that COD removal increases as current density increases from 5 to 10 mA cm$^{-2}$. It can be noticed that increasing the current density beyond 10 mA cm$^{-2}$ did not show any significant improvement in the percentage COD removal because of unwanted side reactions as the pollutant concentration decreases.

Figure 3 shows the BI analysis of electro coagulation using mild steel as an anode at neutral condition, and current density 10 mA cm$^{-2}$. BI value improved from 0.13 to 0.4 within the process time of 40 min, needed for efficient bio degradation [8]. In electrocoagulation, there is an enhanced action on the COD removal by coagulation and adsorption of the pollutants by the dissolved anode, which in turn enhances the BOD to COD ratio, i.e., biodegradability index.

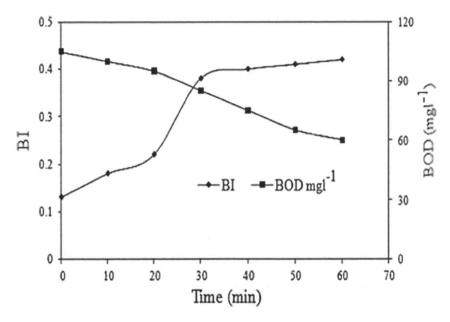

**FIGURE 3:** Variation of BOD and BI with electrolysis time; anode: mild steel, pH 7, current density 10 mA cm$^{-2}$

## 10.3.2 ELECTRO-OXIDATION

The complex mechanism of electrochemical oxidation of wastewater involves the coupling of electron transfer reaction with a dissociate chemisorption step. Basically, two types of oxidative mechanism may occur at the anode; oxidation occurs at the electrode surface in the case of an anode with high electrocatalytic activity, called direct electrolysis; in the other case, oxidation occurs via the surface mediator generated continuously on the anodic surface, called indirect electrolysis (metal oxide electrode). The preferable way for wastewater treatment is the physisorbed route of oxidation. The organic hydrogen peroxides formed are relatively unstable and decomposes which lead to molecular breakdown and the formation of

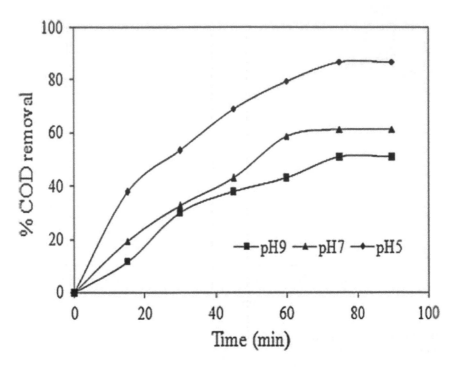

**FIGURE 4:** Effect of pH on % COD removal with electrolysis time; anode: graphite; current density: 10 mA cm$^{-2}$; supporting electrolyte: 5 g l$^{-1}$

subsequent intermediates with lower carbon numbers [9]. This leads to an improvement in the biodegradability index of the wastewater and can be subjected to biological treatment efficiently.

For electro-oxidation method, three parameters such as current density, pH and supporting electrolyte were taken into consideration to determine their effects in COD reduction. Figure 4 explains the influence of pH on COD removal. It has been noticed that treatment at acidic condition gives maximum COD removal up to 86.5 % compared to neutral and basic conditions. Lower pH facilitates the formation of hydroxyl radicals and the organic material in the wastewater can be easily oxidized [10].

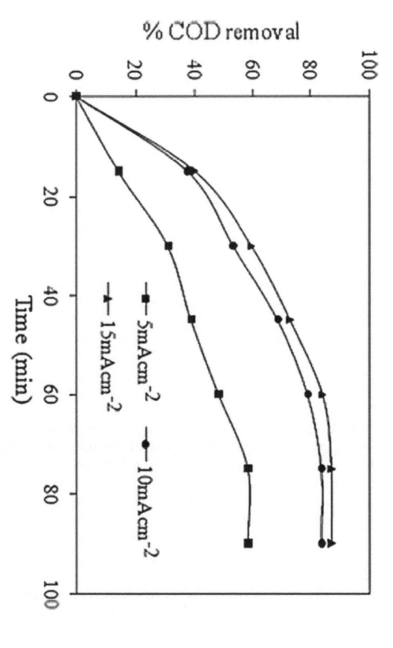

**FIGURE 5:** Effect of current density on percentage COD removal with electrolysis time; anode: graphite; pH 5; supporting electrolyte: 5 g l⁻¹

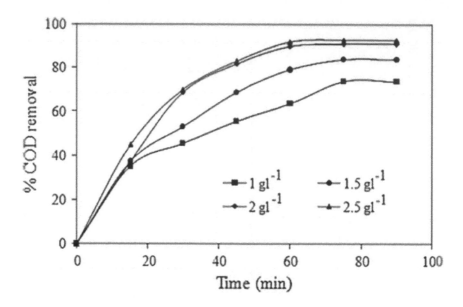

**FIGURE 6:** Effect of supporting electrolyte on percentage COD removal with electrolysis time; anode: graphite; pH 5; current density: 10 mA cm$^{-2}$

From the Fig. 5, it was clear that COD removal increases as current density increases from 5 to 10 mA cm$^{-2}$. It has been noticed that increasing the current density beyond 10 mA cm$^{-2}$ did not show any significant improvement in the percentage COD removal. At higher current densities, the efficiency of COD removal decreases after certain period of time. This is due to the fact that at higher current density, temperature increases and also organic pollutant concentration is lower which leads to side reactions.

From the Fig. 6, it can be understood that efficiency of COD removal increases with increase in supporting electrolyte concentration from 5 to 10 g l$^{-1}$. At higher concentrations, intermediates were formed which do not contribute to COD removal. Also, as pollutant concentration decreases Cl$^-$ radicals in the effluent combine with remaining organic compounds and form complexes, this increases COD of the wastewater [11].

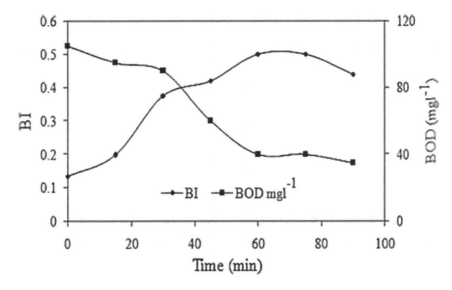

**FIGURE 7:** Variation of BOD and BI with electrolysis time; anode: graphite, pH 4.5, current density 12.5 mA cm$^{-2}$, supporting electrolyte: 7.5 g l$^{-1}$

Figure 7 BI analysis has been made for optimal condition. BI has been improved from 0.13 to 0.4 within the process time of 35 min. After certain period of time, the proportion of degrading organic pollutants is decreased by oxidizing agents, resulting in increased value of BI. The larger molecules in the wastewater are broken down into smaller molecules by the oxidative mechanism and hence facilitate the better action of microorganism on to the organic molecules by biodegradation.

## 10.3.3 PHOTOCHEMICAL REACTION

In photochemical reaction, $H_2O_2$ and ferrous sulphate were added together for oxidation. Advance oxidation process (AOPs) are characterized by the efficient production of hydroxide radicals. The simplified primary

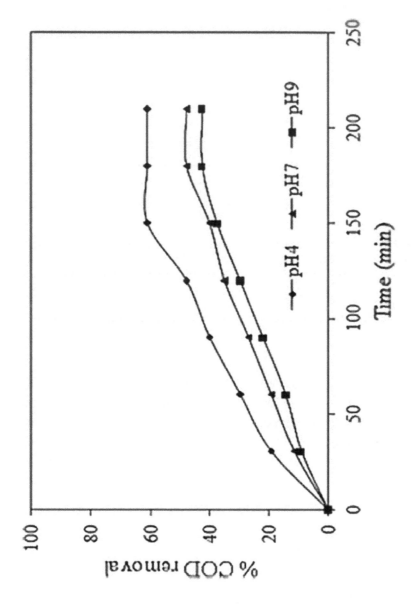

**FIGURE 8:** Effect of pH on percentage COD removal with electrolysis time; $H_2O_2$: 110 mg $l^{-1}$; $FeSO_4$: 50 mg $l^{-1}$

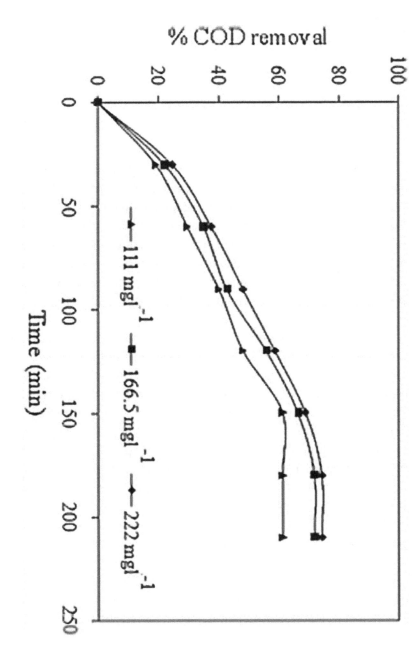

**FIGURE 9:** Percentage COD removal with electrolysis time; pH: 4; FeSO$_4$: 50 mg l$^{-1}$

reactions of photochemical $H_2O_2/Fe^{2+}$-based oxidative water treatment process is given below:

$$H_2O_2 + h\upsilon \rightarrow H_2O_2^* \rightarrow 2OH\bullet \tag{2}$$

$$Fe^{2+} + H_2O_2 \rightarrow OH^- + OH\bullet + Fe^{3+} + H_2O_2 \rightarrow Fe^{2+} + H+ + HO_2\bullet \tag{3}$$

$$Fe^{2+} + H_2O_2 \rightarrow OH^- + OH\bullet + Fe^{3+} + H_2O_2 + h\upsilon \rightarrow Fe^{2+} + H^+ + OH\bullet \tag{4}$$

The above reaction sequence shows the production of OH· radical which acts as an oxidant for the oxidative degradation of pollutants in the wastewater. The various operating parameters like pH and the concentration of $H_2O_2$ and $Fe^{2+}$ has been studied. The effect of pH is given in Fig. 8. It has been noticed that treatment at acidic condition gives maximum COD removal up to 61 % compared to neutral and basic conditions. However, at pH greater than 4, the percentage COD removal decreased with increasing pH from 4 to 9. The hydrogen peroxide starts decomposing at higher pH values [12].

The effect of $H_2O_2$ concentration was studied by changing $H_2O_2$ concentration from 111 to 222 mg l⁻¹. It has been noticed from the Fig. 9 that increase in the concentration of $H_2O_2$ increases the percentage COD removal from 111 to 150 mg l⁻¹. After that it does not show significant increase in percentage COD removal. This is due to the fact that at higher concentration of $H_2O_2$, it destroys hydroxyl radicals formed [13].

From the Fig. 10, it can be noticed that percentage COD removal was increased for $Fe^{2+}$ concentration from 25 to 75 mg l⁻¹. At the higher concentrations of ferrous sulphate, $Fe^{2+}$ combines with hydroxyl radicals which lead to side reactions [12]. From the above figures, it could be concluded that maximum percentage COD removal can be obtained at acidic pH 4 with 166.5 mg l⁻¹ $H_2O_2$ and 50 mg l⁻¹ $Fe^{2+}$ concentrations.

BI analyses were made for treated effluent and are given in the Fig. 11. From the above figure, it can be seen that the BI Index reached the value of 0.4 only after the treatment time of 150 min, which shows that it is not as efficient as electrochemical methods in increasing the biodegradability index of the wastewater and is time consuming.

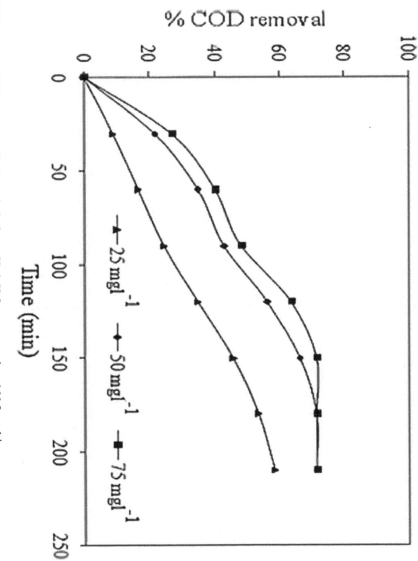

**FIGURE 10:** Percentage COD removal with electrolysis time; pH: 4; $H_2O_2$ concentration: 166.5 mg l$^{-1}$

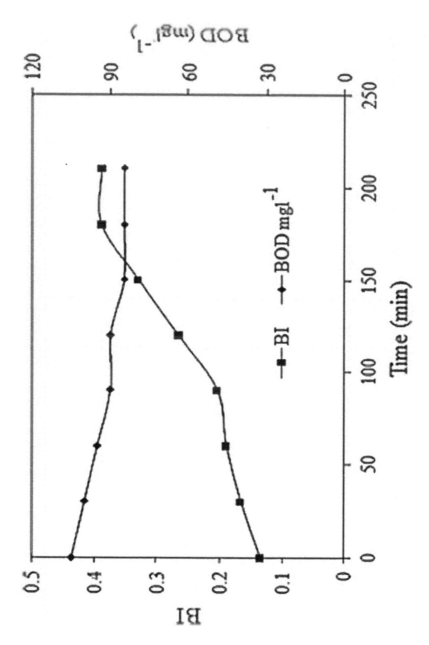

**FIGURE 11:** Variation of BOD and BI improvement with electrolysis time; pH: 4; $H_2O_2$ concentration: 166.5 mg $l^{-1}$; $FeSO_4$: 50 mg $l^{-1}$

## 10.4 CONCLUSION

Though both the electrochemical treatment methods reach the BI of 0.4 at a treatment time of 35–40 min, the electro-oxidation method gives better improvement of BI within the processed time of 35 min without any sludge formation. Then, photochemical method also has disadvantages like usage of expensive UV lamp which is harmful to humans and also it is a very slow process. It can be concluded that the electro-oxidation method is a better option for the improvement of biodegradability index. The treated wastewater now can be further processed by biochemical technique.

## REFERENCES

1. Mahesh S, Prasad B, Mall ID, Mishra IM (2006) Electrochemical degradation of pulp and paper mill wastewater, COD and color removal. Ind Eng Chem Res 45:2830–2839
2. Achoka JD (2002) The efficiency of oxidation ponds at the Kraft pulp and paper mill at Webuye in Kenya. Water Res 36:1203–1212
3. Klavarioti M, Mantzavinos D, Kassinos D (2009) Removal of residual pharmaceuticals from aqueous systems by advanced oxidation processes. Environ Int J 35:402–417
4. Parama Kalyani KS, Balasubramanian N, Srinivasakannan C (2009) Decolorization and COD reduction of paper industrial effluent using electrocoagulation. Chem Eng J 151:97–104
5. Bidhan C, Bag C, Sai M, Sekhar K, Bhattacharya C (2009) Treatment of wastewater containing pyridine released from N, N-Dichlorobis (2,4,6-trichlorophenyl) urea (CC2) plant by advanced oxidation. J Environ Prot Sci 3:34–40
6. Chithra K, Balasubramanian N (2010) Modeling of electrocoagulation through adsorption kinetics. J Model Simul Syst 1:124–130
7. Kobya M, Can OT, Bayramoglu M (2003) Treatment of textile wastewaters by electrocoagulation using iron and aluminum electrodes. J Hazard Mater 100:163–178
8. Chamarro E, MarcoA Esplugas S (2001) Use of fenton reagent to improve organic chemical bio-degradability. Water Res 35:1047–1051
9. Yan L, Ma H, Wang B, Wang Y, Chen Y (2011) Electrochemical treatment of petroleum wastewater with three-dimensional multi-phase electrode. Desalination 276:397–402
10. Ahmed Basha C, Soloman PA, Velan M, Miranda Lima Rose, Balasubramanian N, Slva R (2010) Electrochemical degradation of specialty chemical industry effluent. J Hazard Mater 176:154–164

11. Miled W, Hajsaid A, Roudesli S (2010) Decolorization of high polluted textile wastewater by indirect electrochemical oxidation process. J Text Appar Technol Manag 6:89–97

12. Narendra Kumar B, Anjaneyulu Y, Himabindu V (2011) Comparative studies of degradation of dye intermediate (H-acid) using TiO2/UV/H2O2 and Photo-Fenton process. J Chem Pharma Res 3:718–731

13. Dincer AR, Karakaya N, Gunes N, Gunes Y (2008) Removal of cod from oil recovery industry wastewater by the advanced oxidation processes (aop) based on H2O2. J Glob NEST 10:31–38

# CHAPTER 11

# Removal of Lignin from Wastewater through Electro-Coagulation

RAVI SHANKAR, LOVJEET SINGH, PRASENJIT MONDAL, AND SHRI CHAND

## 11.1 INTRODUCTION

Pulp and paper industry is a water intensive chemical process industry and generates significant quantities of wastewater (20 to 250m3 per ton of pulp produced) containing high concentration of lignin and its derivatives [1]. Some other components such as fatty acid, tannins, resin acid, sulphur compounds, phenol and its derivatives etc. are also present in pulp and paper industry wastewater [2]. The wastewater is generated in five steps in a paper industry such as debarking, pulping, bleaching and alkali extraction, washing and paper production. Amongst the above pollutants, lignin and its derivatives particularly chlorolignin produced during bleaching stage, hold the major share for the development of colour in wastewater. Lignin in solution shows very low BOD: COD ratio (biodegradability index); 600mg/l lignin

*Removal of Lignin from Wastewater through Electro-Coagulation. © 2014 Shankar, Ravi, et al. World Journal of Environmental Engineering 1.2 (2013): 16-20. Used with permission of Science and Education Publishing.*

can contribute around 0-20mg/l of BOD and 750-780mg/l COD value to the wastewater [3]. Due to the low biodegradability index (<0.02) of lignin compounds [3], biochemical methods are not so efficient for the removal of lignin from wastewater as well as for the treatment of pulp and paper industry wastewater. Further, other conventional techniques such as adsorption, chemical oxidation and chemical coagulation require chemicals and produce secondary pollutants. Thus, efforts are on for the development of effective treatment technique for the treatment of pulp and paper industry wastewater around the world. In recent years, electro coagulation (EC) technique has got strong research interest because it produces coagulants in situ by dissolving electrodes in the cell, which helps the removal of the pollutants producing negligible secondary pollutants.

In electro coagulation process, the coagulant is generated in-situ by electro-oxidation of anodes [4]. The generated metal ions due to electro-oxidation of anodes hydrolyze to some extent in water and form soluble monomeric and polymeric hydroxo-metal complexes as shown through Equation.1-4.

Oxidation of anode:

$$M^{(s)} \rightarrow M^{n+} (aq) + ne- \tag{1}$$

Water reduction at cathode:

$$nH_2O\ ne^- \rightarrow n/2H_2 + nOH^- \tag{2}$$

At alkaline conditions:

$$M^{n+} + nOH^- \rightarrow M(OH)_n \tag{3}$$

At acidic conditions:

$$M^{n+} \rightarrow nH_2O \rightarrow M(OH)_n + nH^+ \tag{4}$$

These hydroxo-metal complexes act as coagulant and the whole process occurs through the following steps [4]:

1. Anode dissolution
2. Formation of OH- ions and H2 at the cathode
3. Electrolytic reactions at electrode surfaces
4. Adsorption of colloidal pollutants on coagulant
5. Removal by sedimentation or flotation

Electro-coagulation cannot remove materials that do not form precipitate such as sodium and potassium. If a contaminant cannot form flocks or cannot flocculate, the process will not work out. Therefore the contaminants such as benzene, toluene or similar organic compounds cannot be removed. However, lignin, the macro molecule composed of three monomers namely p-coumaryl alcohol, coniferyl alcohol, and sinapyl alcohol or its derivatives can be removed by electro coagulation.

Some papers are available in literature on the treatment of pulp and paper industry wastewater using electro coagulation process [5, 6, 7, 8, 9]. In most of these papers colour, COD and BOD removal has been considered, however, these papers do not give insight on the removal of COD originated due to lignin compounds. Further, hardly any literature is available, which describes the removal of lignin compounds form wastewater [2].

In the present work the removal of lignin from synthetic waste water by using electro-coagulation with aluminum as a sacrificial electrode in a batch reactor has been described in terms of removal of COD. In the present case the BOD value of the synthetic solution was ~ 26mg/l (below the permissible limit) thus, it was not considered as a parameter for study. Effect of various parameters such as current density, pH, NaCl concentration and treatment time on the removal of COD from synthetic wastewater has been investigated to determine the most suitable process conditions for maximum removal of COD (lignin). A central composite design (CCD) has been used to design the experiment conditions for developing mathematical models to correlate the removal efficiency with the process variables and also to study the interactive influences of parameters on the removal efficiency of pollutants.

## 11.2 MATERIALS AND METHODS

The aluminum used as electrode was sourced from local market in Roorkee, India. The lignin was obtained from Sigma Aldrich, USA and all other chemicals used in the present study were of A. R. grade and purchased from Himedia, Mumbai, India.

The synthetic solution was prepared by mixing 1852g of lignin in one litre double distilled water followed by stirring at 5000RPM for 10 minutes. The characteristic of the synthetic solution is shown in Table 1.

### 11.2.1 PRETREATMENT AND CHARACTERIZATION OF ELECTRODE

The electrode plates were cleaned manually by abrasion with sandpaper followed by further cleaning with 15% HCl for cleaning and washing with distilled water prior to its use. The electrodes were dried for half an hour in an oven and weighted.

TABLE 1: Characteristics of synthetic solution.

| Parameters | Values |
|------------|--------|
| Lignin(mg/l) | 1852 |
| COD (mg/l) | 2500 |
| BOD (mg/l) | 26 |
| pH | 9.28 |

## 11.2.2 BATCH ELECTRO-COAGULATION STUDIES FOR SYNTHETIC SAMPLE

Each 1.5 l of lignin solution was taken in the batch reactor and desired amount of NaCl was added. Then pH of the solution was adjusted using 0.1M HCl and 0.1M NaOH solution. The experimental conditions (current density, pH, NaCl concentration and treatment time) were decided on the basis of a design of experiment as shown in Table 2.

The schematic diagram of the experimental setup is shown in Figure 1. Four aluminum plates, 64cm2 each, were positioned in reactor at distance 1cm from each other. The magnetic stirrer and DC power supply were kept at desired value. The constant current condition was obtained by adjusting the knob of rheostat. After completion of experiment, the sample was centrifuged for 30min at 10000RPM and COD was measured according to closed reflux, colorimetric method [9]. COD value was determined by COD analyzer (SN 09/17443 LOVIBOND Spectrophotometer) after adding desired chemicals and digestion period of 2h in COD Reactor (ET 125SC LOVIBOND).

Lignin content was measured using Folin phenol reagent by UV-spectrophotometer (UV-1800, shimadzu, Japan). BOD5 was measured using BOD analyzer (AL606, Germany) as per standard method [10] using municipal wastewater activated sludge collected from Jagjitpur, Haridwar, as a seed. The experimental data were used to regress mathematical expressions correlating removal efficiency of lignin with the process parameters and to find out a suitable expression through ANOVA using MINITAB software. Error on lignin removal was computed as per Equation.5.

$$Percentage\ error\ (\%) = \frac{(EV - MV)}{EV} \times 100 \qquad (5)$$

Here, EV and MV are the experimental and modeled value.

**TABLE 2:** Characteristics of synthetic solution.

| Experiment Number | A | B | C | D |
|---|---|---|---|---|
| 1 | 70 | 7 | 0.75 | 45 |
| 2 | 130 | 7 | 0.75 | 45 |
| 3 | 70 | 11 | 0.75 | 45 |
| 4 | 130 | 11 | 0.75 | 45 |
| 5 | 70 | 7 | 2.25 | 45 |
| 6 | 130 | 7 | 2.25 | 45 |
| 7 | 70 | 11 | 2.25 | 45 |
| 8 | 130 | 11 | 2.25 | 45 |
| 9 | 70 | 7 | 0.75 | 75 |
| 10 | 130 | 7 | 0.75 | 75 |
| 11 | 70 | 11 | 0.75 | 75 |
| 12 | 130 | 11 | 0.75 | 75 |
| 13 | 70 | 7 | 2.25 | 75 |
| 14 | 130 | 7 | 2.25 | 75 |
| 15 | 70 | 11 | 2.25 | 75 |
| 16 | 130 | 11 | 2.25 | 75 |
| 17 | 40 | 9 | 1.5 | 60 |
| 18 | 160 | 9 | 1.5 | 60 |
| 19 | 100 | 5 | 1.5 | 60 |
| 20 | 100 | 13 | 1.5 | 60 |
| 21 | 100 | 9 | 0 | 60 |
| 22 | 100 | 9 | 3 | 60 |
| 23 | 100 | 9 | 1.5 | 30 |
| 24 | 100 | 9 | 1.5 | 90 |
| 25 | 100 | 9 | 1.5 | 60 |

A:current density (A/m$^2$), B: pH, C: NaCl concentration (mg/l), D:time(min)

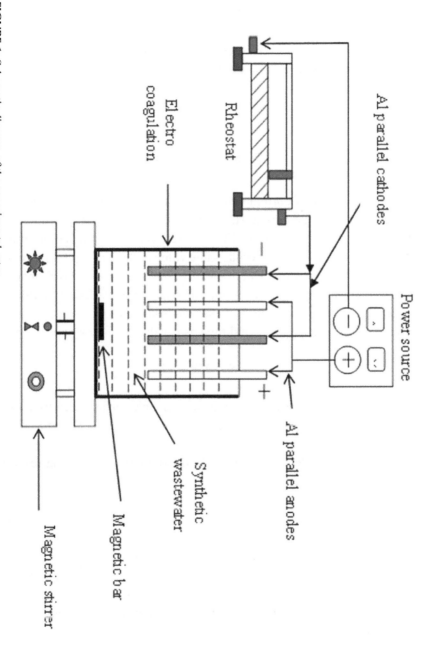

**FIGURE 1:** Schematic diagram of the experimental setup.

## 11.3 RESULT AND DISCUSSION

Effect of various parameters such as current density, pH, NaCl concentration and treatment time on the removal of lignin/COD from synthetic wastewater, model development and interaction effects of process parameters and the validation of the model is described below.

### 11.3.1 EFFECTS OF PROCESS PARAMETERS

A combined effect of current density and initial pH on COD removal is shown in Fig.2. In the present study the COD is generated only due to the lignin in the synthetic solution. Thus, the removal of lignin is proportional to the COD removal, which results same value of % removal for lignin and COD. From Fig. 2 it is evident that COD removal increases with increase in current density and decreases with increase in pH value. According to Faraday law, the the amount of anode materials that dissolve in solution increase with current density and time as per the following expression.

$$m = Mjt \div ZF$$

Where m = amount of ion produced per unit surface area by current density j passed for a duration time t

Z= number of electron involved in the oxidation/ reduction reaction, for alminium(Al), Z=3.

M = atomic weight of material, for Al, M = 26.98g/mol and F is Faraday constant= 96486C/mol.

Thus, at higher current density, higher dissolution of electrode with higher rate of formation of alminium hydroxides results in higher pollutant removal efficiency via co- precipitation and sweep coagulation. However, higher current density also increases solution pH, which reduces the

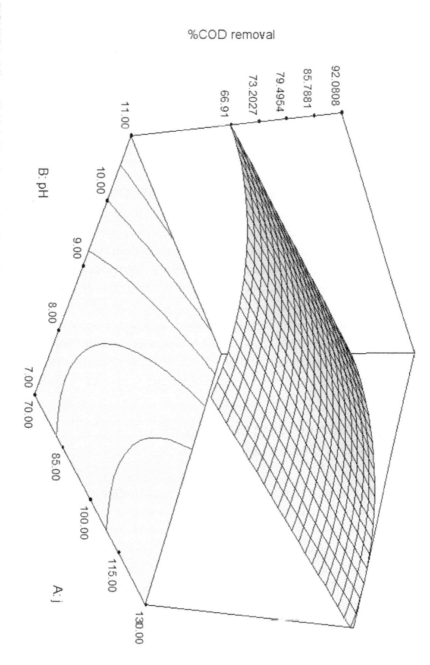

removal efficiency. Under the experimental conditions the most suitable value of j was found as $100A/m^2$.

The effects of pH on the COD (lignin) removal can be explained on the basis of aqueous phase chemistry of lignin and the hydroxo-metal complexes produced by the oxidation of anodic material. The results reveal that at pH ~7.6, the removal efficiency is maximum.

## 11.3.2 EFFECT OF NACL CONCENTRATION AND PH

The effect of NaCl concentration and pH on the COD (lignin) removal is shown in Figure 3. From Figure 3 it seems that the removal of COD decreases with the increase in NaCl concentration as well as increase in pH. The addition of sodium chloride increases the conductivity of the wastewater and thus energy consumption becomes low with addition of electrolyte. Sodium chloride was selected because of low toxicity, reasonable cost and the fact that NaCl prevents the organic matter to attach on the surface of anode, which can create inhibition. However, a higher dosage of it induces overconsumption of electrodes, which increases the aluminum content in sludge; consequently removal efficiency of COD (lignin) is reduced. The most suitable value of NaCl concentration is found to be 0.75mg/l.

## 11.3.3 EFFECT OF TREATMENT TIME
## AND NACL CONCENTRATION

The effect of treatment time and NaCl concentration on the removal efficiency of COD (lignin) is shown in Figure 4.

From Figure 4 it is clear that with increse in treatment time the COD (lignin) removal increases. The increase in removal with time can be explained on the basis of Faraday's law which shows that the amount of Al ion and their flocks generated are propotional to time. It has been observed that a treatment time of 75min is sufficient to remove ~ 95% of the COD (lignin) from the solution under the experimental conditions.

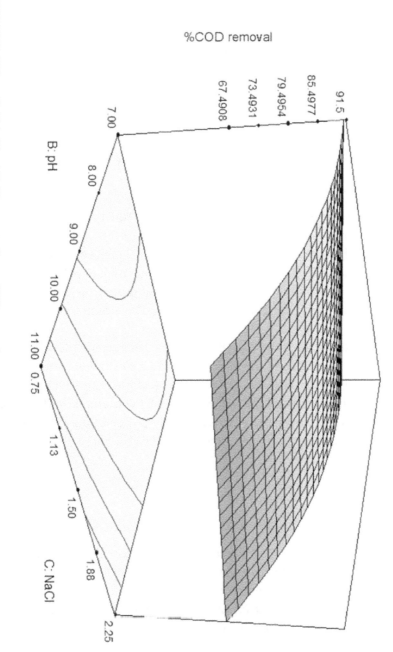

**FIGURE 3:** Effect of initial pH and NaCl concentration on COD removal

## 11.3.4 MODEL DEVELOPEMNT

The proposed model equation to correlate COD (lignin) removal with input process parameters in coded factors is shown in Equation.6.

$$
\begin{aligned}
\text{COD(liglin) removal} = \\
61.1 + 0.183A - 2.46B + 11.0C \\
+ 0.834D + 0.0207AB - 2.76BC \\
- 0.032CD - 0.00604DA + 0.000176ABCD
\end{aligned}
\tag{6}
$$

$(R^2 = 0.85, \text{ F value } 18.99, \text{ P value} < 0.05)$

where, A= Current density $(A/m^2)$, B = pH, C = NaCl concentration (mg/l), D = time (min)

From the above equation it seems that NaCl concentrttaion highly influences the COD (lignin) removal and the interation of pH and NaCl concentration is also highest amongst the above process parameters. This is in accordance with the effect of pH and NaCl concentration as discussed above.

From the main effect of the process parameters (developed from MINITAB software) on the % removal of COD (P) from waste water as shown in Figure.5, it is evident that the % removal of COD (lignin) decreases with increase in the pH(B) and NaCl concentration (C), whereas it increases with increase in current density(A) and time (D). The above model is able to predict the % removal of COD (lignin) with the error limit of +9 to -7%.

## 11.4 CONCLUSIONS

Removal of COD (lignin) increases with increase in current density and treatment time and decreases with increase in pH and NaCl concentration. Under the most suitable conditions i.e, current density: $100A/m^2$, pH: 7.6, NaCl concentration: 0.75mg/l and treatment time: 75min, around 95% removal of COD ( lignin) is achieved from the synsthetic solution. The empirical model gives prediction on the % removal of COD (lignin) with + 9 to -7% error limit. The present process can reduce COD value from

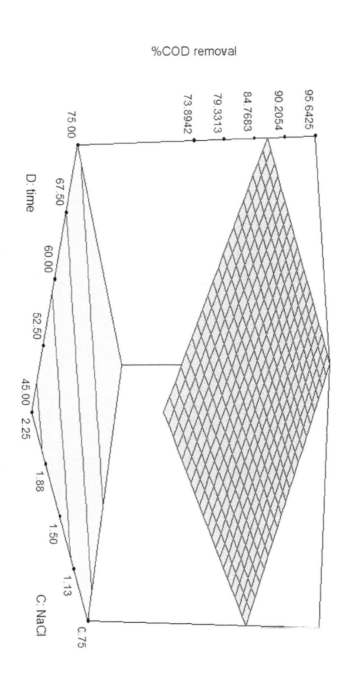

**FIGURE 4:** Effect of time and NaCl concentration on COD removal

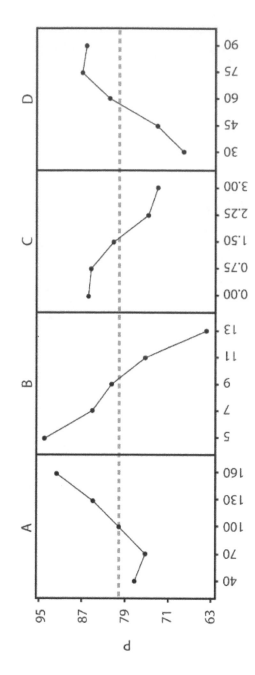

**FIGURE 5:** Main effects of process variables on the % removal of COD (obtained from MINITAB software).

~2500mg/l to ~ 125mg/l after treatment, which is below the maximum permissible limit of COD in discharge water.

## REFERENCES

1.  Garg, A., Mishra, I.M., and Chand, S., "Catalytic wet oxidation of the pretreated synthetic pulp and paper mill effluent under moderate conditions", Chemosphere, 66(9), 1799-1805. January 2007.
2.  Pokhrel, D. and Viraraghavan,T., "Treatment of pulp and paper mill wastewater—a review," Science of the Total Environment, 333 (1-3). 37-58. October 2004.
3.  Kindsigo, M. and Kallas, J., "Degradation of lignins by wet oxidation: model water solutions," Proc. Estonian Acad. Sci. Chem., 55 (3). 132-144. 2006.
4.  Ceylan, Z., Cana, B.Z. and Kocakerim, M.M., "Boron removal from aqueous solutions by activated carbon impregnated with salicylic acid," Journal of Hazardous Materials, 152(1). 415-422. March 2008.
5.  Katal, R. and Pahlavanzadeh, H, "Influence of different combinations of aluminum and iron electrode on electro coagulation efficiency: Application to the treatment of paper mill wastewater," Desalination, 265 (1-3). 199-205. January 2011.
6.  Kalyani, K.S.P., Balasubramanian, N. and Srinivasakannan, C, "Decolorization and COD reduction of paper industrial effluent using electro-coagulation," Chemical Engineering Journal, 151 (1-3). 97-104. August 2009.
7.  Mahesh, S., Prasad, B., Mall, I.D., and Mishra, I.M., "Electrochemical Degradation of Pulp and Paper Mill Wastewater. Part 1. COD and Color Removal," Industrial & Engineering Chemistry Research, 45 (8). 2830-2839. March 2006.
8.  Sridhar, R., Sivakumar, V., Immanuel, V.P., and Maran, J.P, "Treatment of pulp and paper industry bleaching effluent by electro-coagulant process," Journal of Hazardous Materials, 186 (2-3), 1495-1502, February 2011.
9.  Vepsalainen, M., "Removal of toxic pollutants from pulp mill effluents by electro coagulation," Separation and Purification Technology, 81 (2). 141-150. September 2011.
10. Cleceri, L.S., Greenberg, A.E., and Eaton A.D, Standard methods for the examination of water and wastewater, American Public Health Association, USA, Washington DC, 1998, 20th ed.

# CHAPTER 12

# Treatment of Wastewater Effluents from Paper-Recycling Plants by Coagulation Process and Optimization of Treatment Conditions with Response Surface Methodology

NOUSHIN BIRJANDI, HABIBOLLAH YOUNESI, AND NADER BAHRAMIFAR

## 12.1 INTRODUCTION

The pulp and paper mills are among the most important industries in the world, but also some of the biggest polluting agents, discharging a variety of pollutants such as gaseous, liquid and solid wastes into the environment. The pollution of water bodies is of major global concern, because these industries generate large volumes of wastewater, viz. about 80 m3 of wastewater for each ton of pulp produced (Oguz and Keskinler 2008). More than 250 chemicals produced at different stages of paper production have been identified in the effluents (Thompson et al. 2001). According to research reports, samples with biodegradability index ($BOD_5/COD$) smaller than 0.3 are not appropriate for biological degradation (Helble et

*Treatment of Wastewater Effluents from Paper-Recycling Plants by Coagulation Process and Optimization of Treatment Conditions with Response Surface Methodology.* © *Noushin Birjandi, Habibollah Younesi, and Nader Bahramifar 2014.* International Journal of Industrial Chemistry *March 2014, 5:4. Published with SpringerLink.com (http://dx.doi.org/10.1007/s13201-014-0231-5). Creative Commons Attribution License.*

al. 1999), as for complete biodegradation the effluent must present an index of at least 0.40 (Chamarro et al. 2001).

Paper mill wastewater can cause considerable damage to the recipient
waters due to the high chemical oxygen demand (COD) and high toxicity.
To minimize the impact of effluents on the environment, several treatment technologies have been employed, although little is known on their
efficiency to eliminate the toxicity attributed to the presence of organic
compounds. This is mainly due to the fact that it is the type of paper mills
(packaging, recycling, kraft) and the water system configuration that determine the COD, toxicity and organic load of the effluent. Therefore, the
design and efficiency of wastewater treatments will vary from mill to mill
(Latorre et al. 2007). Thompson et al. (2001) reviewed the different types
of treatment of pulp and paper mill effluents and indicated effective processes to minimize the discharge of wastewater to the environment. Coagulation is one of the most used water effluent treatments. It employs a
cationic metal as a coagulant agent which usually promotes water hydrolysis and the formation of hydrophobic hydroxide compounds with different
charges, depending on the solution pH. It may also lead to the formation of
polymeric compounds. The coagulants interact with colloidal materials by
either charge neutralization or adsorption, leading to coagulation usually
followed by sedimentation (Stephenson and Duff 1996). The coagulation
effectiveness and cost depend on the coagulant type and concentration,
solution pH, ionic strength as well as on both concentration and nature
of the organic residues in the effluent (Afzal et al. 2008). The response
surface methodology (RSM) is a statistical technique for designing experiments, building models, evaluating the effects of several factors, searching
optimum conditions for desirable responses and reducing the number of
experiments (Wang et al. 2007). RSM has been proposed to determine
the influence of individual factors and their interactive influence. It uses
an experimental design such as the central composite design (CCD) in
order to fit a modeling by the least squares technique, and the adequacy
of the proposed model is then revealed using the diagnostic checking tests
(Anderson-Cook et al. 2009).

The main objective of this work was to optimize the coagulation process and investigate the interactive effects of the experimental factors, viz.

the coagulant dosages, pH and initial value of COD as process parameters. For this purpose, paper-recycling wastewater was selected as the target to be treated by the coagulation process, optimizing it by RSM under Design-Expert software. The quadratic models obtained were used in a constrained optimization to achieve optimum process conditions for maximum removal efficiencies of turbidity and COD and the lowest sludge volume index (SVI) of the paper-recycling wastewater.

## 12.2 MATERIALS AND METHODS

### 12.2.1 ANALYTICAL METHODS

The paper-recycling wastewater used in this work was taken from Afrang Paper Manufacture Ltd., Iran. The wastewater samples were characterized according to standard methods for the examination of water and wastewater (APHA 1998) and the results are given in Table 1. Alum and PACl (Merck) were used as coagulants. The initial solution pH of the paper-recycling wastewater was adjusted using 1.0 mol/l sodium hydroxide and 0.5 mol/l sulfuric acid solution. A calorimetric method with closed reflux was developed for the measurement of COD (the Plaintest system, photometer 8000, England) at 600 nm, which was used to measure the absorbance of the COD samples. A pH meter (the Waterproof CyberScan PC 300, Singapore) was used to measure the solution pH. Turbidity was measured by a turbidity meter (the CyberScan TB 1000 Eutech Instruments, Singapore). The SVI was measured using a 1,000 ml Imhoff cone (APHA 1998).

### 12.2.2 EXPERIMENTAL PROCEDURE

The coagulation experiments were performed with 100 ml paper-recycling wastewater, adjusting the initial COD and pH at different values. In a typical run, the paper-recycling wastewater was diluted with tap water to adjust its COD concentration to the desired value. Then different

**TABLE 1:** Chemical and physical characteristics of the wastewater used.

| Characteristics | Range | Mean |
|---|---|---|
| Color | Brown | – |
| pH | 6.32–7.60 | 6.5 |
| COD (mg/l) | 3,348–3,765 | 3,523 |
| $BOD_5$ (mg/l) | 870–974 | 940 |
| TS (mg/l) | 3,067–3,307 | 3,260 |
| TSS (mg/l) | 245–267 | 260 |
| Ash (mg/l) | 1,206–1,345 | 1,320 |
| Turbidity (TNU) | 855–880 | 873 |
| Phosphate (mg/l) | 68–76 | 73 |
| Sulfate (mg/l) | 45–55 | 50 |
| Nitrate (mg/l) | 0.89–1.12 | 1.03 |
| Nitrite (mg/l) | 0.07–0.13 | 0.10 |
| Chloride (mg/l) | 9.8–11.1 | 10.7 |

concentrations of coagulants were progressively added. The samples were stirred at 50 rpm (Velp scientific magnetic stirrer) for 2 min to completely dissolve the coagulant. This was followed by a further slow mixing for 20 min at 40 rpm. The flocks formed were allowed to settle for 30 min. After settling, the turbidity and final COD of the supernatant were determined. The remaining portion of the treated wastewater samples was used to determine the SVI.

## 12.2.3 EXPERIMENTAL DESIGN FOR OPTIMIZATION OF PARAMETERS

The RSM was applied for developing, improving and optimizing the processes and to evaluate the relative significance of several affecting factors even in the presence of complex interactions. In the present study, the De-sign- Expert 7.0 (State-Ease, Inc., Minneapolis, MN, USA) software was used for regression and graphical analyses of the obtained data. RSM, as a robust design technology based on the central composite design (CCD), could be applied to the modeling and analysis of multiple parameters. The experimental design had with four different factors: solution pH, initial COD value, alum dosage, PACl dosage. Each of the parameters was coded at five levels: $-\alpha$, $-1$, $0$, $+1$ and $+\alpha$. A CCD involves a two-level factorial (k 2) design, 2 k axial points (denoted by $\pm \alpha$), and n c center points. In this study, the goal was to create a full factorial design at the k = 4 design, completely randomized, in one block with six center points and the axial distance $\alpha = 2$ for a rotatable design. The range and level of the variables in coded units from the RSM studies are given in Table 2. For statistical calculations, the variable X i was coded as x i, according to the following equation:

$$x_i = \frac{X_i - X_0}{\Delta X} \qquad (1)$$

where $X_1$ is the uncoded value of the independent variable, i, $X_0$ is the value of $X_1$ at the center point of the investigated area and $\Delta X$ is the step change. Table 3 shows the coded and uncoded variables for the four ex-perimental variables according to the Eq. (1). To obtain the optimum dos-ages, initial COD and pH, three dependent parameters were analyzed as response, final COD and turbidity removal efficiencies and SVI value. The quadratic equation model for predicting the optimal conditions can be expressed according to Eq (2) as follows:

$$Y = \beta_0 + \sum_{i=1}^{k} \beta_i X_i + \sum_{i=1}^{k} \beta_{ii} X_i^2 + \sum_{i=1}^{k-1} \sum_{j=2}^{k} \beta_{ij} X_i X_j + \varepsilon \qquad (2)$$

where (i) is the linear coefficient, (ii) the quadratic coefficient ($\beta$) the regression coefficient, (ij) the interaction coefficient, (k) the number of

**TABLE 2:** Experimental ranges and levels of the independent variables.

| Independent variables | Range and level | | | | |
|---|---|---|---|---|---|
| | −α | −1 | 0 | +1 | +α |
| COD (A) | 250 | 750 | 1,250 | 1,750 | 2,250 |
| Alum dosage [mg/l (B)] | 200 | 650 | 1,100 | 1,550 | 2,000 |
| PACl dosage [mg/l (C)] | 55 | 420 | 785 | 1,150 | 1,515 |
| pH (D) | 2 | 4.5 | 7 | 9.5 | 12 |

factors studied and optimized in the experiment and ($\varepsilon$) the random error. The experiments were carried out according to the experimental data sheet (both coded and uncoded design matrices) shown in Table 3, which also lists the response values, i.e., COD removal, turbidity removal and SVI value. The results are further analyzed using Design-Expert Software. The relationship between the four controllable factors (Alum and PACl dosages, pH and initial COD) and the three important operating parameters (COD removal, turbidity removal and SVI) for coagulation process was studied.

## 12.3 RESULTS AND DISCUSSION

### 12.3.1 RSM APPROACH OF OPTIMIZATION OF COD AND TURBIDITY REMOVAL AND SVI VALUE

The statistical significance of the quadratic model was evaluated by the analysis of variance (ANOVA), as presented in Table 4. The empirical relationship between the COD removal (Y COD), turbidity removal (Y Tur) and SVI value (Y SVI) and the four test variables in coded units

**TABLE 3:** Experimental ranges and levels of the independent variables.

| Run | Uncoded (coded) values | | | | Responses | | |
|---|---|---|---|---|---|---|---|
| | A | B | C | D | Final COD removal (%) | Turbidity (%) | SVI (ml/g) |
| 1 | 750 (−1) | 1,550 (+1) | 1,150 (+1) | 4.5 (−1) | 84 | 69 | 244 |
| 2 | 2,250 (+α) | 1,100 (0) | 785 (0) | 7 (0) | 73 | 91 | 295 |
| 3 | 1,750 (+1) | 650 (−1) | 420 (−1) | 4.5 (−1) | 79 | 77 | 400 |
| 4 | 250 (−α) | 1,100 (0) | 785 (0) | 7 (0) | 67 | 65 | 850 |
| 5 | 750 (−1) | 650 (−1) | 1,150 (+1) | 9.5 (+1) | 80 | 46 | 994 |
| 6 | 1,250 (0) | 1,100 (0) | 785 (0) | 7 (0) | 82 | 81 | 394 |
| 7 | 1,250 (0) | 1,100 (0) | 55 (−α) | 7 (0) | 84 | 83 | 805 |
| 8 | 1,750 (+1) | 650 (−1) | 420 (−1) | 9.5 (+1) | 79 | 84 | 803 |
| 9 | 1,250 (0) | 1,100 (0) | 1,515 (+α) | 7 (0) | 88 | 80 | 205 |
| 10 | 1,750 (+1) | 650 (−1) | 1,150 (+1) | 9.5 (+1) | 84 | 78 | 662 |
| 11 | 1,750 (+1) | 1,550 (+1) | 1,150 (+1) | 4.5 (−1) | 82 | 82 | 712 |
| 12 | 1,750 (+1) | 650 (−1) | 1,150 (+1) | 4.5 (−1) | 80 | 77 | 354 |
| 13 | 750 (−1) | 650 (−1) | 420 (−1) | 4.5 (−1) | 82 | 69 | 988 |
| 14 | 750 (−1) | 1,550 (+1) | 420 (−1) | 9.5 (+1) | 72 | 72 | 889 |
| 15 | 1,750 (+1) | 1,550 (+1) | 1,150 (+1) | 9.5 (+1) | 86 | 87 | 186 |
| 16 | 1,250 (0) | 1,100 (0) | 785 (0) | 7 (0) | 83 | 77 | 371 |
| 17 | 1,250 (0) | 1,100 (0) | 785 (0) | 2 (−α) | 88 | 60 | 50 |
| 18 | 1,250 (0) | 200 (−α) | 785 (0) | 7 (0) | 80 | 58 | 1,690 |
| 19 | 750 (−1) | 650 (−1) | 1,150 (+1) | 4.5 (−1) | 84 | 60 | 214 |

**TABLE 3:** CONTINUED.

| | | | | | | | |
|---|---|---|---|---|---|---|---|
| 20 | 750 (−1) | 650 (−1) | 420 (−1) | 9.5 (+1) | 75 | 68 | 1,970 |
| 21 | 1,250 (0) | 2,000 (+α) | 785 (0) | 7 (0) | 82 | 72 | 892 |
| 22 | 1,250 (0) | 1,100 (0) | 785 (0) | 7 (0) | 82 | 79 | 355 |
| 23 | 1,750 (+1) | 1,550 (+1) | 420 (−1) | 4.5 (−1) | 82 | 67 | 525 |
| 24 | 750 (−1) | 1,550 (+1) | 1,150 + 1 | 9.5 (+1) | 75 | 67 | 297 |
| 25 | 750 (−1) | 1,550 (+1) | 420 (−1) | 4.5 (−1) | 85 | 68 | 830 |
| 26 | 1,250 (0) | 1,100 (0) | 785 (0) | 7 (0) | 83 | 76 | 269 |
| 27 | 1,250 (0) | 1,100 (0) | 785 (0) | 7 (0) | 81 | 81 | 295 |
| 28 | 1,250 (0) | 1,100 (0) | 785 (0) | 12 (+α) | 83 | 69 | 347 |
| 29 | 1,250 (0) | 1,100 (0) | 785 (0 | 7 (0) | 81 | 79 | 384 |
| 30 | 1,750 (+1) | 1,550 (+1) | 420 (−1) | 9.5 (+1) | 81 | 84 | 93 |

by the application of RSM is given by quadratic models, as given in Table 4. The adequacy of the model was tested through lack-of-fit, P values and F values (Vohra and Satyanarayana 2002). The P values for lack-of-fit were 0.4028, 0.6499 and 0.54488 for the removal of COD and turbidity and SVI value, respectively, which indicates that it is not statistically significant as is desirable. The insignificant value of lack-of-fit (>0.05) shows that the model is valid for the present study (Ziagova et al. 2007). The F values of 44.50, 54.61 and 156.90 were for COD removal and turbidity removal efficiency and SVI value, respectively, and P values were significant (P < 0.05) (Table 4). This means that the model terms are statistically significant and can predict response satisfactorily. The fit of the model was checked by determining the coefficient ($R^2$). In this study, the $R^2$ values obtained were 0.9765, 0.9880 and 0.9932 for the removal of COD and turbidity and SVI value, respectively. The closer the $R^2$ is to 1, the stronger is the model and the better it predicts the response. The value of the adjusted determination of the coefficient (adjusted $R^2$ = 0. 9545, 0.9628 and 0.9869) for COD and turbidity removal efficiency and SVI is also high, showing the high significance of the model. The values of the predicted $R^2$ at 0.8926, 0.9214 and 0.9708 obtained for the COD and turbidity removal efficiency and SVI value, respectively, were also high, supporting the considerable significance of the model. At the same time, the relatively low values of the coefficient of variation ($CV_{COD}$ = 1.22, $CV_{Tur}$ = 2.57 and $CV_{SVI}$ = 8.41) indicate a good precision and reliability of the experiments (Nakano and Jutan 1994; Collins et al. 2009). Adequate precision (AP) is a measure of the range in the predicted response relative to its associated error. Its desirable value is 4 or more. Adequate precision compares the range of the predicted values at the design points with the average prediction error. A ratio greater than 4 indicates an adequate model discrimination (Aghamohammadi et al. 2007; Ghafari et al. 2009). Usually it is necessary to check the fitted model to ensure that it provides an adequate approximation with the real system. Unless the model shows an adequate fit, proceeding with an investigation and optimization of the fitted response surface is likely to give poor or misleading results. Diagnostic plots such as the predicted versus actual values help us to judge the satisfactoriness of the model.

**TABLE 4:** The developed models and ANOVA results for the response parameters.

| Models in terms of coded values of parameters | F value | P value | F-LOF | P-LOF | R² | Adj. R² | Pred. R² | CV (%) | AP |
|---|---|---|---|---|---|---|---|---|---|
| $Y_{COD} = + 82.00 + 1.17A + 0.33B + 1.17C − 1.50D + 0.88AB + 0.13AC + 2.50AD − 0.37BC − 0.75BD + 1.00CD − 3.00A^2 − 0.25B^2 + 1.00C^2 + 0.88D^2$ | 44.50 | <0,0,001 | 1.31 | 0.4028 | 0.9765 | 0.9545 | 0.8926 | 1.22 | 29.97 |
| $Y_{Tur} = + 78.83 + 7.04A + 2.71B − 1.21C + 1.46D − 1.81AB + 2.94AC + 2.69AD + 3.19BC + 1.94BD − 2.31CD − 0.22A^2 − 3.47B^2 + 0.66C^2 − 3.59D^2$ | 54.61 | <0.0001 | 0.79 | 0.6499 | 0.9808 | 0.9628 | 0.9214 | 2.57 | 33.03 |
| $Y_{SVI} = + 344.67 − 158.38A − 175.21B − 168.13C + 92.54D + 75.19AB + 188.81AC − 132.56AD + 64.94BC − 207.44BD − 24.81CD + 55.81A^2 + 235.47B^2 + 38.97C^2 − 37.66D^2$ | 156.90 | 0.0001 | 0.98 | 0.5448 | 0.9932 0.9708 | 0.9869 | 0.9708 | 8.41 | 53.558 |

From this table, Y is pollutant removal (response) in percentage, A, B, C and D are the coded values of the tests variables initial COD, alum dose, PACl dose and solution pH, respectively, AP adequate precision, LOF lack-of-fit, Adj. R² adjusted R², Pred. R² predicted R².

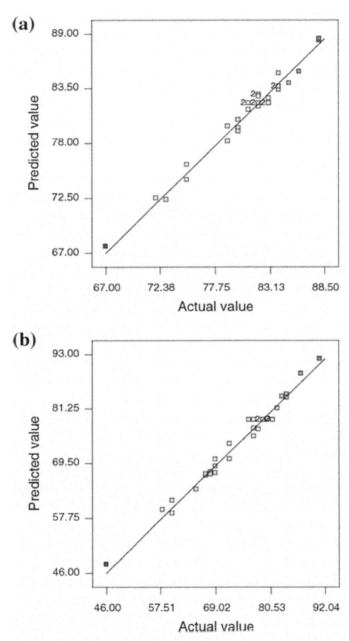

**FIGURE 1:** Design-expert plot; predicted vs. actual values plot for **a** final COD removal ($R^2 = 0.9765$), **b** turbidity removal ($R^2 = 0.9808$) and **c** SVI ($R^2 = 0.9932$). The colors changes from blue to red in the response value indicating 0 to 100 % range

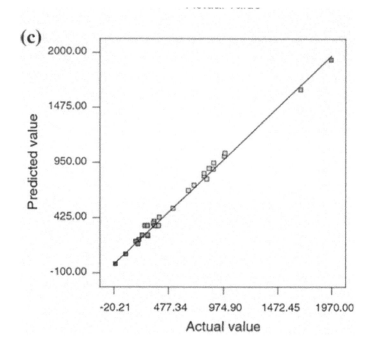

**FIGURE 1:** CONTINUED.

Figure 1 shows the plots for predicted versus actual values of the parameters for removal of the COD (a) and reduction of the turbidity (b) and SVI value (c). These plots indicate an adequate agreement between the uncoded data and the ones obtained from the models. By applying the diagnostic plots such as the ones for predicted versus actual values, the model adequacy can be judged. The regression equation after the ANOVA gave the level of COD and turbidity removal as a function of four factors, i.e., the dosages of alum and PACl, initial solution pH and initial COD. After applying multiple regression analysis on the experimental data, the results of the CCD design were fitted with a second-order full polynomial equation.

## 12.3.2 EFFECT OF DIFFERENT ALUM AND PACL DOSAGE ON COD AND TURBIDITY REDUCTION

In coagulation processes, three factors, viz. kind of an inorganic co-agulant, the coagulant dosage and the pH, play an important role in determining the coagulation process efficiency in wastewater. In treatments using inorganic coagulants, the optimum pH range in which metal hydroxide precipitates should be determined (Aguilar et al. 2002). Different coagulants affect different degrees of destabilization. The higher the valence of the counter ion, the more is its destabilizing effect and the less is the dose needed for coagulation (Birjandi et al. 2013). At high dose of metal ions (coagulant), a sufficient degree of oversaturation occurs to produce a rapid precipitation of a large quantity of metal hydroxide, enmeshing the colloidal particles which are termed as sweep floc (Hu et al. 2011).The effect of different levels of the alum and PACl doses on COD and turbidity removal efficiency can be predicted from the 3D plot, while the other two variables are at its middle level, as shown in Fig. 2. Figure 2a clearly shows that the COD removal efficiency increases with increasing coagulant dosages. The maximum COD removal (88.26 %) is shown with 657 mg/l dosage of alum and 1,500 mg/l dosage of PACl. Thus, for significant removal of COD, high doses of coagulants are needed. This is due to the presence of large amounts of organic matter in the wastewater and interaction with coagulants that can cause the suspended solids in the effluent to be oxidized, coagulated and then be settled as sludge; this process will reduce COD (Kumar et al. 2009). In Fig. 2b the effects of PACl dosage on turbidity reduction are not significant within the dosage range studied. The effect of increasing the PACl dosage only reveals a minor impact on the reduction-removal efficiency of turbidity. It can be seen that the maximum removal of turbidity (84.8 %) was obtained with approximately 853 mg/l dosage of alum and 53 mg/l dosage of PACl. The addition of PACl in the effluent and its mixing create proper coagulation condition and the flocs generated are denser than water, hastening the settling of the flocs. The PACl, having multivalent aluminum ions, neutralize the particle charges and the hydrolyzed aluminum flocs enmesh the colloids and drive to settle at high COD (Hu et al. 2011).

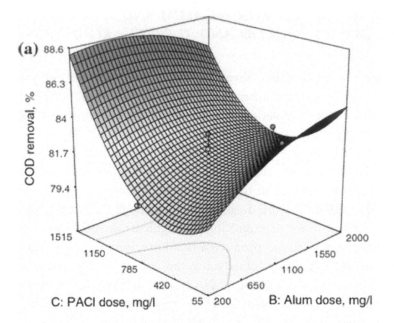

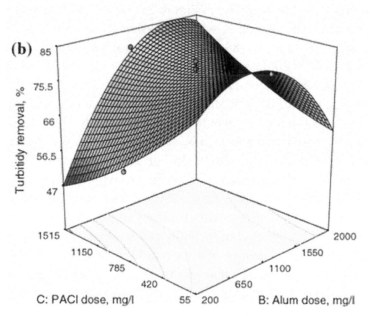

**FIGURE 2:** Effect of coagulants doses on **a** COD removal, **b** turbidity reduction, while other two variables at its middle level. The colors changes from blue to red in the response value indicating 0 to 100 % range.

The use of alum alone is not a perfect method for treatment paper mill wastewater and it is best to investigate with the other methods treatment and coagulants that can be provide necessary standards for this effluent. Therefore, in paper-recycling wastewater treatment use of alum coupled with PACl could be the best results (Stephenson and Duff 1996).

## 12.3.3 EFFECT OF SOLUTION PH AND ALUM DOSE ON COD AND TURBIDITY REDUCTION

Changes in coagulants species and charge of the target compound can result due to variation in pH of the liquid media. In addition to governing the coagulant speciation, the media pH influences the extent of dissociation of the trace organic contaminants, and can result in compound-specific removal performance during application of a certain type of coagulant (Birjandi et al. 2013). The addition of metal coagulants depresses the wastewater pH to a lower value. The decrease in pH after the addition of coagulant may be due to the several hydrolytic reactions, which are taking place during coagulation, forming multivalent charged hydrous oxide species and generating $H_3O^+$ ion during each step, thus reducing the pH value (Kumar et al. 2011).

It has also been reported that the coagulant addition depresses pH to highly acidic levels, as the coagulant dose is highly correlated with pH (Chaudhari et al. 2007). It is supposed that improvement of flocculation pH may reduce the alum dose needs for the optimization of the process. Figure 3 presents the effect of different levels of the pH and the alum dose on the removal of COD and turbidity when other two variables at its middle level. Both the turbidity reduction and the COD removal efficiency increase with an decreasing pH adjustment till reach their highest value at the optimal pH, between which the reduction efficiency values start to decrease. This confirms the findings of (Ahmad et al. 2008). Figure 3a shows that the maximum COD removal (89.67 %) can be obtained at pH = 2.0, but Fig. 3b shows that the maximum turbidity reduction efficiency (79.73 %) occurs at pH = 7.82. It is well known that the pH of the solution affects

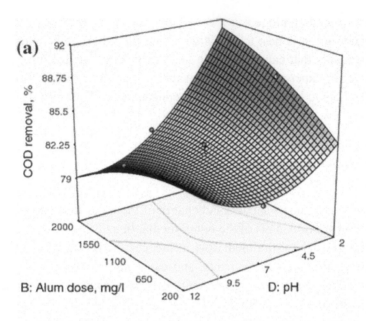

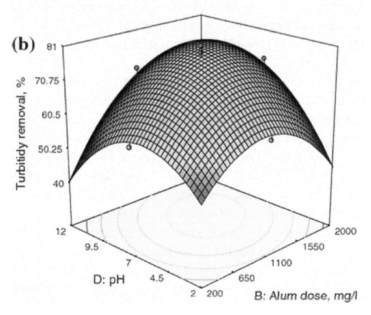

**FIGURE 3:** Effect of pH and alum dose on **a** COD removal and **b** turbidity reduction, while other two variables at its middle level. The colors changes from blue to red in the response value indicating 0 to 100 % range

the hydroxyl radical generating capacity (Zayas et al. 2007). According to Petala et al. (2006), at a lower pH and with a lower coagulant dosage, the only mechanism for destabilization of particles is through charge neutralization. At a low pH, because the aggregates are small in size, the mechanism of colloidal destabilization is mainly charge neutralization. It can be said that the highest range of pH exists between 6.0 and 8.0, beyond which the effluent quality deteriorates. The most important reasons for such behaviours are: 1) at low pH, the presence of monomeric aluminum species [$Al^{3+}$, $Al(OH)^{2+}$ and $Al(OH)_2{}^+$] causes the anionic particles to neutral and deposition of pollutants settlement process is best performed due to the formation positive metal complexes that help to build flocs with the negative organic pollutants in wastewater, 2) with increase in pH, the concentration of dissolved aluminum is reduced by the formation of uncharged metal hydroxide [$Al(OH)3$] that leads to rapid precipitation and 3) with further increase in pH, the species $Al(OH)_4{}^-$ is dominant which reduces coagulation effects and no coagulation occurs anymore (Ahmad et al. 2008).

## 12.3.4 EFFECT OF INITIAL COD AND PH
## ON FINAL COD AND TURBIDITY REMOVAL EFFICIENCY

The experimental runs were conducted with five different initial COD values. The effect of different levels of the initial COD concentration and the pH on final COD and turbidity removal efficiency can be predicted from the 3D plot, while the other two variables are at its middle level, as shown in Fig. 4. Figure 4a shows that the maximum value of 89.67 % for COD removal is found with an initial COD of 940 mg/l and that the COD removal efficiency decreases with a high initial COD value and with COD of 250 mg/l a COD removal efficiency of 58 % is achieved. In contrast, the results obtained on turbidity show that high initial COD values result in high reduction efficiency (Fig. 4b). With COD of 250 mg/l (as the COD low limit value introduced to software), turbidity reduction efficiency of 42 % is achieved, while the maximum value of 95 % reduction is found at initial COD of 2,250 mg/l.

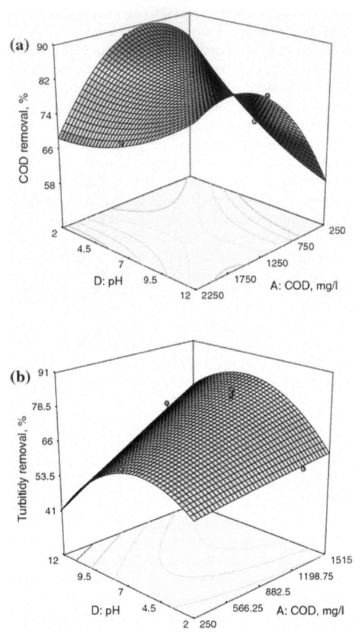

**FIGURE 4:** Effect of initial COD and pH on **a** COD removal and **b** turbidity reduction, while other two variables at its middle level. The colors changes from blue to red in the response value indicating 0 to 100 % range

## 12.3.5 SLUDGE VOLUME INDEX (SVI)

The sludge produced in physical–chemical treatments is due to the organic matter and total solids in suspension that are removed and the compounds formed with the coagulants used, since practically all of the latter become part of the sludge solids. In general, the amount and characteristics of the sludge produced during the coagulation process depend on the coagulants used and on the operating conditions. To observe the volume and settling characteristics, the SVI was determined. The SVI in coagulation process is generally governed by three factors: high polymer effect, osmotic pressure effect and hydration effect (Wang et al. 2007). Figure 5a presents the effect of different levels of the alum and PACl dosages on the SVI value when other two variables at its middle level. From this figure, the coagulant dosages of alum coupled with PACl have a significant interaction on the SVI, as shown in Fig. 5a. The curvilinear profile obtained for SVI is in accordance to the quadratic model. The SVI decreases significantly with coagulant dosage and pH, but is affected mostly by the coagulant dosage. The best condition of the minimum SVI (140 ml/g) is observed with dosage of 1,140 mg/l alum and of 1,495 mg/l PACl. Under strong acid conditions (e.g. pH < 4.5) and low ranges of COD, the aluminum ions are mostly in the form of $Al^{3+}$, which is highly effective for decreasing the osmotic pressure and hydration effects. Figure 5b presents the effect of different levels of the pH and the initial COD concentration on the SVI value when the other two variables are at its middle level. All of these factors might be responsible for the much lower SVI under acid conditions (Fig. 5b). These results, observed from the response surface plot, are in good agreement with the fitted model for SVI obtained earlier. Figure 5c presents the effect of different levels of the PACl dose and the initial COD concentration on the SVI value when other two variables at its middle level. From this figure, the results obtained on SVI show that high initial COD values result in high SVI value. With COD of 750 mg/l (as the COD low limit value introduced to software), SVI value of about 340 ml/g is achieved, while the maximum value of 375 ml/g is found at initial COD of 1,750 mg/l.

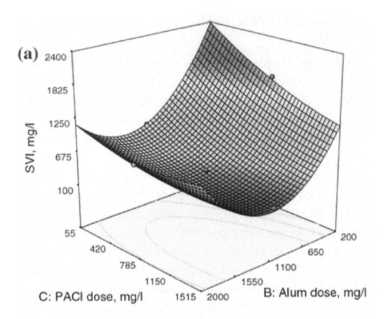

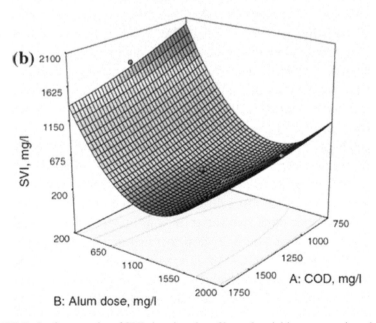

**FIGURE 5:** Surface graphs of SVI showing the effect of variables: **a** coagulant dosages, **b** alum dose and COD and **c** PACl dose and COD. The colors changes from blue to red in the response value indicating 0 to 100 % range.

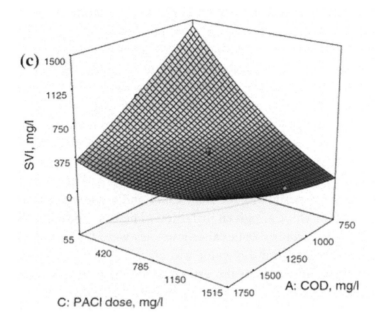

**FIGURE 5:** CONTINUED.

## 12.3.6 OPTIMIZATION USING THE DESIRABILITY FUNCTION

In the numerical optimization, we choose the desirable goal for each factor and response from the menu. The possible goals are: maximize, minimize, target, within range, none (for responses only) and set to an exact value (factors only). A minimum and a maximum level must be provided for each parameter included. A weight can be assigned to each goal to adjust the shape of its particular desirability function. The goals are combined into an overall desirability function, an objective function that ranges from zero outside of the limits to one at the goal. The selected program in numerical optimization seeks to maximize this function. The goal seeking begins at a random starting point and proceeds up the steepest slope to a maximum. There may be two or more maximums because of curvatures in the response surface and their combination into the desirability function. By starting from several points in the design space, chances improve for finding the best local maximum. A multiple response method was applied for optimizing any combination for four goals, namely, the initial solution pH, alum dosage, PACl dosage and initial COD. The numerical optimization found a point that maximizes the desirability function. The importance of each goal was changed in relation to the other goals. Optimized conditions and results of experiments for final COD and turbidity removal efficiency and SVI value are given in Table 5. The best optimum value for final COD and turbidity removal efficiencies were 80.02 and 83.23 %, respectively, and desirability value was close to 1. The obtained values of desirability showed that the estimated function may represent the experimental model and desirable conditions. In Table 6, the results of this study are compared with the results of other researchers; although wastewater and coagulants differ, but the use of alum coupled with PACl lead to remove most pollutants from wastewater is corollary to our methods. In our previous published paper, the purpose was aimed to examine the efficiency of alum and PACl (coagulants) in combination with a cationic polyacrylamide (C-PAMs, i.e., chemfloc 1510c and chemfloc 3876 as flocculating agents) in the removal of COD and turbidity from paper-recycling wastewater. The results demonstrated that the maximum amounts of 40 mg/l coagulant dosage and 4.5 mg/l flocculant dosage at pH 4.5 were required to give 92 % removal of turbidity, 97 % removal of COD and 80 ml/g value of SVI. However, the best coagulant and flocculant were alum and chemfloc 3876 at a dose of 41 and 7.52 mg/l, respectively, at pH of 6.85. In these conditions, the highest removal of COD and turbidity and lowest value of SVI were found to be 91.30 and 95.82 % and 12 ml/g, respectively.

**TABLE 5:** Optimized conditions and results of experiments for final COD and turbidity removal efficiencies and SVI value.

| No. | Variable | | | | Predicted value | | | Desirability value | Experimental value | | |
|---|---|---|---|---|---|---|---|---|---|---|---|
| | COD (mg/l) | Alum dose (mg/l) | PACl dose (mg/l) | pH | COD removal (%) | Turbidity removal (%) | SVI (ml/g) | | COD removal (%) | Turbidity removal (%) | SVI (ml/g) |
| 1 | 1,750 | 1,550 | 1,314 | 9.5 | 85.32 | 88.36 | 132.24 | 0.936 | 80.02 | 83.23 | 140.00 |
| 2 | 1,736 | 1,320 | 724 | 9.5 | 81.87 | 86.62 | 67.42 | 0.862 | 82.30 | 80.12 | 65.00 |
| 3 | 1,750 | 1,240 | 490 | 9.5 | 81.05 | 88.85 | 104.20 | 0.847 | 85.15 | 82.52 | 112.64 |

**TABLE 6:** Comparison of the present finding with other researchers for the same used coagulants.

| Wastewater | pH | COD (mg/l) | Turbidity (NTU) | Coagulant | Dose of coagulant (g/l) | COD removal (%) | Turbidity removal (%) | References |
|---|---|---|---|---|---|---|---|---|
| Leachate | 8.2–8.5 | 1,794–2,094 | 268–502 | Alum | 9.5 | 60.8 | 88.9 | Ghafari et al. (2009) |
| | | | | PACl | 2 | 46.0 | 94.9 | |
| Pulp mill | 9.5 | 1,303 | 10 | FeCl$_3$ | 0.08 | 58.0 | 75.0 | Rodrigues et al. (2008) |
| Vinasse | 8.4 | 8,525 | 4,600 | FeCl$_3$ | 20 | 99.2 | 84.0 | Zayas et al. (2007) |
| Paper-recycling | 6.5 | 3,523 | 872.5 | Alum | 1.55 | 82.0 | 83.2 | Present study |
| | | | | PACl | 1.31 | | | |

## 12.4 CONCLUSIONS

Coagulation process is one of the simple and common physical–chemical methods, advocated to be used for paper-recycling wastewater treatment. Although there are many types of coagulants available to treat water and wastewater, opting the most effective coagulant for a particular wastewater is important. The treatment of pulp and paper mill wastewater using alum coupled with PACl as coagulants enhanced the reduction removal of turbidity and COD and produced a lower volume of sludge compared to the results obtained when the coagulants were used alone. A desirable functional approach was used to obtain a compromise between three different responses, i.e., COD, turbidity removal and SVI. The optimum conditions obtained were with 1,550 mg/l alum coupled with 1,314 mg/l PACl at pH 9.5, with 80.02 % of COD removal, SVI of 140 ml/g and 83.23 % of turbidity removal. The results showed good agreement between the experimental and model predictions.

## REFERENCES

1. Afzal M, Shabir G, Hussain I, Khalid ZM (2008) Paper and board mill effluent treatment with the combined biological-coagulation-filtration pilot scale reactor. Bioresour Technol 99(15):7383–7387
2. Aghamohammadi N, HbA Aziz, Isa MH, Zinatizadeh AA (2007) Powdered activated carbon augmented activated sludge process for treatment of semi-aerobic landfill leachate using response surface methodology. Bioresour Technol 98(18):3570–3578
3. Aguilar MI, Sáez J, Lloréns M, Soler A, Ortuño JF (2002) Nutrient removal and sludge production in the coagulation-flocculation process. Water Res 36(11):2910–2919
4. Ahmad AL, Wong SS, Teng TT, Zuhairi A (2008) Improvement of alum and PACl coagulation by polyacrylamides (PAMs) for the treatment of pulp and paper mill wastewater. Chem Eng J 137(3):510–517
5. Anderson-Cook CM, Borror CM, Montgomery DC (2009) Response surface design evaluation and comparison. J Stat Plan Inference 139(2):629–641
6. APHA (1998) Standard methods for the examination of water and wastewater, 20th edn. American Public Health Association, Washington, DC
7. Birjandi N, Younesi H, Bahramifar N, Ghafari S, Zinatizadeh AA, Sethupathi S (2013) Optimization of coagulation-flocculation treatment on paper-recycling wastewater: application of response surface methodology. J Environ Sci Health, Part A 48(12):1573–1582

8.  Chamarro E, Marco A, Esplugas S (2001) Use of fenton reagent to improve organic chemical biodegradability. Water Res 35(4):1047–1051

9.  Chaudhari P, Mishra I, Chand S (2007) Treatment of biodigester effluent with energy recovery using various inorganic flocculant. Colloids Surf A Physicochem Eng Aspects 296:238–247

10. Collins LM, Dziak JJ, Li R (2009) Design of experiments with multiple independent variables: a resource management perspective on complete and reduced factorial designs. Psychol Methods 14(3):202–224

11. Ghafari S, Aziz HA, Isa MH, Zinatizadeh AA (2009) Application of response surface methodology (RSM) to optimize coagulation-flocculation treatment of leachate using poly-aluminum chloride (PAC) and alum. J Hazard Mater 163(2–3):650–656

12. Helble A, Schlayer W, Liechti P-A, Jenny R, Möbius CH (1999) Advanced effluent treatment in the pulp and paper industry with a combined process of ozonation and fixed bed biofilm reactors. Water Sci Technol 40(11–12):343–350

13. Hu XJ, Wang JS, Liu YG, Li X, Zeng GM, Bao ZL, Zeng XX, Chen AW, Long F (2011) Adsorption of chromium (VI) by ethylenediamine-modified cross-linked magnetic chitosan resin: Isotherms, kinetics and thermodynamics. J Hazard Mater 185 (1):306–314. doi:10.1016/j.jhazmat.2010.09.034

14. Kumar R, Singh R, Kumar N, Bishnoi K, Bishnoi NR (2009) Response surface methodology approach for optimization of biosorption process for removal of Cr(VI), Ni (II) and Zn (II) ions by immobilized bacterial biomass sp. Bacillus brevis. Chem Eng J 146 (3):401–407. doi:10.1016/j.cej.2008.06.020

15. Kumar P, Teng TT, Chand S, Wasewar KL (2011) Treatment of paper and pulp mill effluent by coagulation. Int J Civil Environ Eng 3(3):222–227

16. Latorre A, Malmqvist A, Lacorte S, Welander T, Barceló D (2007) Evaluation of the treatment efficiencies of paper mill whitewaters in terms of organic composition and toxicity. Environ Pollut 147(3):648–655

17. Nakano E, Jutan A (1994) Application of response surface methodology in controller fine-tuning. ISA Trans 33(4):353–366

18. Oguz E, Keskinler B (2008) Removal of colour and COD from synthetic textile wastewaters using O3, PAC, H2O2 and H2CO3. J Hazard Mater 151(2–3):753–760

19. Petala M, Tsiridis V, Samaras P, Zouboulis A, Sakellaropoulos GP (2006) Wastewater reclamation by advanced treatment of secondary effluents. Desalination 195(1–3):109–118

20. Rodrigues AC, Boroski M, Shimada NS, Garcia JC, Nozaki J, Hioka N (2008) Treatment of paper pulp and paper mill wastewater by coagulation-flocculation followed by heterogeneous photocatalysis. J Photochem Photobiol A Chem 194(1):1–10

21. Stephenson RJ, Duff SJB (1996) Coagulation and precipitation of a mechanical pulping effluent–I. Removal of carbon, colour and turbidity. Water Res 30(4):781–792

22. Thompson G, Swain J, Kay M, Forster CF (2001) The treatment of pulp and paper mill effluent: a review. Bioresour Technol 77(3):275–286

23. Vohra A, Satyanarayana T (2002) Statistical optimization of the medium components by response surface methodology to enhance phytase production by Pichia anomala. Process Biochem 37(9):999–1004

24. Wang J-P, Chen Y-Z, Ge X-W, Yu H-Q (2007) Optimization of coagulation-flocculation process for a paper-recycling wastewater treatment using response surface methodology. Colloids Surf Physicochem Eng Aspects 302(1–3):204–210
25. Zayas T, Rómero V, Salgado L, Meraz M, Morales U (2007) Applicability of coagulation/flocculation and electrochemical processes to the purification of biologically treated vinasse effluent. Sep Purif Technol 57(2):270–276
26. Ziagova M, Dimitriadis G, Aslanidou D, Papaioannou X, Litopoulou Tzannetaki E, Liakopoulou-Kyriakides M (2007) Comparative study of Cd(II) and Cr(VI) biosorption on Staphylococcus xylosus and Pseudomonas sp. in single and binary mixtures. Bioresour Technol 98(15):2859–2865

# PART V

# CONCLUSION

# CHAPTER 13

# Life Cycle Assessment Applications in Wastewater Treatment

NURDAN BUYUKKAMACI

## 13.1 INTRODUCTION

Life Cycle Assessment (LCA) is a technique to assess all potential environmental impacts of any action and covers the entire life of the product; the study begins with raw material acquisition through production, use, and disposal [1]. A life-cycle uses the "cradle-to-grave" approach, and identifies energy use, material input, and waste generated from the acquisition of raw materials to the final disposal of the product. The facility at which this is conducted will benefit from the identification of the areas where environmental improvements can be made. Its principles and requirements are defined by ISO 14000 series standards [2,3] and consist of four main activities: goal and scope (ISO 14040), inventory analysis (ISO 14044), impact assessment (ISO 14044), and interpretation (ISO 14044).

Although LCA is generally used for the production phase, it is also applied to the global environmental analysis of a wastewater treatment operation. The environmental assessment of wastewater treatment technologies

*Life Cycle Assessment Applications in Wastewater Treatment.* © *2013 Buyukkamaci N.* Journal of Pollution Effects & Control *1:104 (doi: 10.4172/2375-4397.1000104). Creative Commons Attribution License.*

has been realized with the LCA technique. Corominas et al. [4] reported that LCA applied to wastewater treatment is a field with 17 years of experience and more than 40 studies have been published in international peer-reviewed journals using an array of databases, boundary conditions, and impact assessment methods for interpreting the results since 1990s.

The wastewater treatment plants help us to protect the environment, but in contrast to their main commissioned purpose, they can damage the environment through energy consumption, greenhouse gas emission, the utilization of chemicals, and some toxic material outcomes. Among them energy consumption is one of the most important issues in treatment plants. In wastewater treatment plants, huge amounts of energy are usually consumed and the amount of energy needed for operations varies depending on effluent characteristics, treatment technology, required effluent quality, and plant size [5]. Energy is required at every stage of the treatment plant, including pumping, mixing, heating, and aeration. The wastewater treatment plants should be designed and operated considering the amount of energy consumption. The one of the largest energy consumers in the treatment plant are aeration equipment [6]. Conventional activated sludge or aerated lagoon wastewater treatment processes can efficiently remove organic pollutants, but the operation of such systems is cost and energy intensive, mainly due to the aeration and sludge treatment associated processes [7]. Anaerobic treatment is generally a more environmentally friendly treatment technology than aerobic treatment due to its low solids generation rate, low electrical energy requirements and the production of a usable biogas [8]. Even though anaerobic degradation results in greenhouse gas emissions, these emissions and the energy requirements of the treatment systems could be reduced by the generation energy from biogas. So, when deciding on the treatment and disposal options in wastewater treatment plant design, all impacts should be taken into consideration. For example, water reuse is a very popular approach to protect natural water sources and it is thought to be an environmentally friendly application. However, in water reuse applications, generally high quality water is required, and therefore water reclamation facilities generally include additional advanced treatment technologies, which can consume large amounts of energy. In this case, LCA can be used for the assessment and the comparison between the different techniques.

Although most of the LCA studies focused on the energy consuming [9-12] there are also some studies evaluating greenhouse gas emissions [13,14], toxicity [15,16], and eutrophication [15,17]. Muñoz et al. [15] use LCA for the comparison of two solar-driven advanced oxidation processes, namely heterogeneous semiconductor photo catalysis and homogeneous photo-Fenton, both coupled to biological treatment, are carried out in order to identify the environmentally preferable alternative to treat industrial wastewaters containing non-biodegradable priority hazardous substances. In LCA, global warming, ozone depletion, human toxicity, freshwater aquatic toxicity, photochemical ozone formation, acidification, eutrophication, energy consumption, and land use were taken into consideration. Depending on the assessment results, an industrial wastewater treatment plant based on heterogeneous photo catalysis involves a higher environmental impact than the photo-Fenton alternative. Zhang et al. [9] applied LCA to illuminate the benefits of a wastewater treatment and reuse project in China. Energy consumption was used as the sole parameter for quantitative evaluation of the project. As a result of the LCA analysis of the case project in Xi'an, China, it was revealed that the life cycle benefits gained from treated wastewater reuse greatly surpassed the life cycle energy consumption for the tertiary treatment. Other researchers from China investigated the environmental impacts associated with the treatment of wastewater in a wastewater treatment plant (WWTP) in Kunshan, China by Life Cycle Assessment. SimaPro 7.0 software was used in this study and the construction phase, operation, and maintenance phase; sludge landfilling and the transportation of chemicals to the WWTP were all taken into consideration. The LCA results of Kunshan WWTP taking renewable energy (wind power) as the energy source proposed that enhancing the effluent quality will decrease the environmental impact [18]. Amores et al. [19] empltoyed the Life Cycle Assessment methodology to carry out an environmental analysis of every stage of the urban water cycle in Tarragona, a Mediterranean city of Spain, taking into account a water supply system of a city considering water abstraction, potable water treatment, distribution network, wastewater treatment, reclaimed water, and desalination. They compared three scenarios: 1) current situation, 2) using reclaimed water and using desalination plants, 3) reclaimed water to supply water during a drought. In all three scenarios the main source of impact was the energy consumed through the collection and intermediate pumping of freshwater.

As indicated by Corominas et al. [4], different methodologies have been applied at LCA studies and there is variability in the definition of the functional unit and the system boundaries, the selection of the impact assessment methodology and the procedure followed for interpreting the results. It is important to use same functional unit to compare the alternative options. The functional unit is defined by the service provided by the system being studied and further shaped by the goal of the study [20,21] and selected depending upon the aims of the study [22]. The volume unit of treated wastewater and one dry ton of sludge are commonly used for wastewater treatment and sludge handling and disposal processes, respectively [9,13,22,23].

The system boundaries determine which unit processes shall be included within the LCA [2]. In the inventory analysis, allows of materials and energy across the system boundary are quantified [21]. Some of the studies cover only the operation phases [11] the others cover the construction phase [9,24] or the disposal and transportation phase [13,18]. In LCA studies, so many kinds of life cycle assessment methods, such as Eco Indicator 99, EDIP 96, EPS and Ecopoints 97 [22], and commercial software, such as SimaPro [18,25], Umberto 5.5 [26], and Gabi [27], has been used.

## 13.2 CONCLUSION

In conclusion, the benefits and harms of each application should be investigated in detail and among the alternative treatment methods, the most environmentally friendly treatment options, especially the least energy consuming techniques, should be selected and applied. In order to design and construct the most appropriate wastewater treatment plants, it would be useful to use the LCA approach and different environmental impacts of wastewater treatment plants should be identified by the LCA method. Further analysis should be carried out on life cycle impact assessment of wastewater treatment techniques.

# REFERENCES

1. Matjaz P (2004) Environmental impact and life cycle assessment of heating and airconditioning systems, a simplified case study. Energ Buildings 36: 1021-1027.
2. ISO 14040:2006 Environmental management - Life cycle assessment - Principles and framework.
3. ISO 14044:2006 Environmental management - Life cycle assessment - Requirements and guidelines.
4. Corominas L, Foley J, Guest JS, Hospido A, Larsen HF, et al. (2013) Life Cycle Assessment Applied to Wastewater Treatment: State of the art 47: 5480-5492.
5. Hernández-Sancho F, Molinos-Senante M, Sala-Garrido R (2011) Energy efficiency in Spanish wastewater treatment plants: A non-radial DEA approach. Sci Total Environ 409: 2693-2699.
6. Buyukkamaci N, Koken E (2010) Economic evaluation of alternative wastewater treatment plant options for pulp and paper industry. Sci Total Environ 408: 6070-6078.
7. HugginsT, Fallgren PH, Jin S,RenZJ (2013) Energy and Performance Comparison of Microbial Fuel Cell and Conventional Aeration Treating of Wastewater. J Microbial Biochem Technol.
8. Fiss N, Smith S (2007) Green Industrial Wastewater Treatment Utilizing Anaerobic Processes. NC AWWA-WEA 87th Annual Conference, Charlotte, NC.
9. Zhang QH, Wang XC, Xiong JQ, Chen R, Cao B (2010) Application of life cycle assessment for an evaluation of wastewater treatment and reuse project - Case study of Xi'an, China. Bioresource Technol 101: 1421-1425.
10. Remy C, Jekel M (2012) Energy analysis of conventional and source-separation-systems for urban wastewater management using Life Cycle Assessment. Water Sci Technol 65: 22-29.
11. Bravo L,Ferrer I (2011) Life Cycle Assessment of an intensive sewage treatment plant in Barcelona (Spain) with focus on energy aspects. Water Sci Technol 64: 440-447.
12. Tan X, Xu J, Wang S (2011) Application of Life Cycle Analysis (LCA) in evaluating energy consuming of Integrated Oxidation Ditch (IOD). Procedia Environmental Sciences 5: 29-36.
13. Liu B, Wei Q, Zhang B, Bi J (2013) Life cycle GHG emissions of sewage sludge treatment and disposal options in Tai Lake Watershed, China. Sci Total Environ 447:361-369.
14. Ashrafi O, Yerushalmi L, Haghighat F (2013) Greenhouse gas emission by wastewater treatment plants of the pulp and paper industry - Modeling and simulation. Int J Greenhouse Gas Control 17: 462-472.
15. Munoz I, Peral J, Ayllon JA, Malato S, Passarinho P, et al. (2006) Life cycle assessment of a coupled solar photocatalytic-biological process for wastewater treatment. Water Res 40. 3533-3540

16. Igos E, Benetto E, Venditti S, Kohler C, Cornelissen A, et al. (2012) Is it better to remove pharmaceuticals in decentralized or conventional wastewater treatment plants? A life cycle assessment comparison. Sci Total Environ 438: 533-540.

17. Wang X, Liu J, Ren NQ, Duan Z (2012) Environmental profile of typical anaerobic/ anoxic/oxic wastewater treatment systems meeting increasingly stringent treatment standards from a life cycle perspective. Bioresource Technol 126: 31-40.

18. Li Y, Luo X, Huang X, Wang D, Zhang W (2013) Life Cycle Assessment of a municipal wastewater treatment plant: A case study in Suzhou, China. J Clean Prod 57: 221-227.

19. Amores MJ, Meneses M, PasqualinoJ, Antón A, Castells F (2013) Environmental assessment of urban water cycle on Mediterranean conditions by LCA approach. J Clean Prod 43: 84-92.

20. Curran MA (2013) Life Cycle Assessment: a review of the methodology and its application to sustainability. Curr Opin Chem Eng 2: 273-277

21. Burgess AA, Brennan DJ (2001) Application of life cycle assessment to chemical processes. Chem Eng Sci 56: 2589-2604.

22. Renou S, Thomas JS, Aoustin E, Pons MN (2008) Influence of impact assessment methods in wastewater treatment LCA. J Clean Prod 16: 1098-1105.

23. Houillon G, Jolliet O (2005) Life cycle assessment of processes for the treatment of wastewaterurban sludge: energy and global warming analysis. J Clean Prod 13: 287-299.

24. Foley J, Haas D, Hartley K, Lant P (2010) Comprehensive life cycle inventories of alternative wastewater treatment systems. Water Res 44: 1654-1666.

25. Hancock NT, Black ND, Cath TY (2012) A comparative life cycle assessment of hybrid osmotic dilution desalination and established seawater desalination and wastewater reclamation processes. Water Res 46: 1145-1154.

26. Benetto E, Nguyen D, Lohmann T, SchmittB, Schosseler P (2009) Life cycle assessment of ecological sanitation system for small-scale wastewater treatment. Sci Total Environ 407: 1506 - 1516.

27. Tangsubkul N, Parameshwaran K, Lundie S, Fane AG, Waite TD (2006) Environmental life cycle assessment of the microfiltration process. J Membr Sci 284: 214-226.

# AUTHOR NOTES

## CHAPTER 1

### Acknowledgments

This work was supported by the Spanish MCI through the project CTM2010-15682. The authors appreciate help from JM Navarro and M Garcia with the GC/MS analyses.

## CHAPTER 3

### Acknowledgments

The authors gratefully acknowledge the support provided by the Biochemical and Bioenvironmental Engineering Research Center (BBRC) of Sharif University of Technology, Tehran, Iran.

### Author Contributions

MV, have made substantial contribution to conception, design and arrangement of this research project. MS, carried out the analysis and interpretation of data and pre drafting the manuscript. IA as corresponding author, revising the manuscript critically for important intellectual content, final approval of the version to be published; and agree to be accountable for all aspects of the work. All authors read and approved the final manuscript.

### Competing Interests

The authors declare that they have no competing interests.

## CHAPTER 4

### Author Contributions
Conceived and designed the experiments: CP CT FB. Performed the experiments: CP CT FB. Analyzed the data: CP CT FB. Contributed reagents/materials/analysis tools: CP CT FB. Wrote the paper: CP CT FB.

### Competing Interests
The authors have declared that no competing interests exist.

## CHAPTER 5

### Conflict of Interests
The authors do not have a direct financial relationship with the commercial identities mentioned in the paper that might lead to a conflict of interests.

### Acknowledgments
The authors thank the Ministry of Economic Affairs of Republic of China for financial support of this research under Contract no. 101-EC-17-A-10-S1-187.

## CHAPTER 6

### Competing Interests
The authors do not have a direct financial relationship with the commercial identities mentioned in the paper that might lead to a conflict of interests.

### Acknowledgments
The authors are very much thankful to the Laboratory of Photoelectron Spectroscopy (LEFE), Universidade Estadual Paulista "Júlio de Mesquita Filho" (UNESP), Araraquara, São Paulo, Brazil, by the performing the XPS analysis in this research work.

### Author Contributions
Conceived and designed the experiments: MRS. Performed the experiments: MRS. Analyzed the data: MRS RAA CAON BC. Contributed reagents/materials/analysis tools: MRS BC. Wrote the paper: MRS RAA CAON BC.

## CHAPTER 7

### Acknowledgment

This research was supported by the National Science Council, Taiwan, under Grant no. NSC 100-2221-E-562-001-MY3.

## CHAPTER 9

### Competing Interests

The authors declare that there is no conflict of interests regarding the publication of this paper.

### Acknowledgment

Authors acknowledge the enterprise International Rectifiers, Anonymous Society, which through Manuel González, Eng., gave a confident vote to a research group situated in Tijuana to develop a systematic and scientific work that helped in a better operation of their wastewater treatment plant. Authors also thank the Autonomous University of Baja California, Mexico, UABC, that throughout Dr. José Manuel Cornejo-Bravo provide all facilities to use the Zetasizer equipment for zeta potential measurements. Eduardo Alberto López-Maldonado gives his gratitude to the National Council of Science and Technology (CONACYT) in Mexico, for the fellowship received in his Ph. D. studies.

## CHAPTER 11

### Competing Interests

The authors have no competing interests.

## CHAPTER 12

### Acknowledgments

The present research was made possible by the sponsorship and financial support of the Ministry of Science of Iran and the Tarbiat Modares University (TMU). The authors wish also to thank Mrs Haghdoust (Technical

Assistant of Environmental Laboratory) of TMU for her cooperation and Ellen Vuosalo Tavakoli (University of Mazandaran) for the final editing of the English text.

# INDEX